Adrian Caesar was born near Manchester, and has lived in Australia since 1982. He is Associate Professor in English at the University College of NSW, Australian Defence Force Academy. He is the author of several books of literary and cultural criticism and two books of poems.

the white

adrian caesar

MACMILLAN

First published 1999 in Picador by Pan Macmillan Australia Pty Limited

This edition published 2001 by Macmillan
an imprint of Macmillan Publishers Ltd
25 Eccleston Place, London SW1W 9NF
Basingstoke and Oxford
Associated companies throughout the world
www.macmillan.com

ISBN 0 333 90572 5

Copyright © Adrian Caesar 1999

The right of Adrian Caesar to be identified as the
author of this work has been asserted by him in accordance
with the Copyright, Designs and Patents Act 1988.

All rights reserved. No part of this publication may be
reproduced, stored in or introduced into a retrieval system, or
transmitted, in any form, or by any means (electronic, mechanical,
photocopying, recording or otherwise) without the prior written
permission of the publisher. Any person who does any unauthorized
act in relation to this publication may be liable to criminal
prosecution and civil claims for damages.

1 3 5 7 9 8 6 4 2

A CIP catalogue record for this book is available
from the British Library.

Printed and bound in Great Britain by
Mackays of Chatham plc, Chatham, Kent

For Victor

Author's Note

This is not an 'official' biography of Scott and Mawson. Some will say that it is not a biography at all. What I have attempted to write here is an experiment in biography in which I use my extensive researches to imaginatively re-create the last days in the Antarctic journeys of Scott and Mawson, 1911–1913. In using my imagination, I have included material that I cannot 'know' to be true. But I have tried by the narrative method employed to signal to the reader those aspects of the story that are verifiable and those that rely on imaginative interpretation. In other words, instead of trying to conceal the fictional nature of biography, I have chosen to emphasise that aspect. Consequently, the responsibility for the interpretation of the explorers' lives herein is all mine and I claim no final authority for them.

I am very grateful to Alun Thomas, Sir Douglas Mawson's grandson, for his time and willingness to share documents relating to his grandfather with me. I would also like to thank the following for their help in the research of the book: Susan Woodburn at the Barr Smith Library, University of Adelaide; Mark Pharaoh, archivist at Urbrae House, University of Adelaide; Lois Baker, Dr Jennifer McDonell, Susan Cowan and Sar ah Randles of the School of English, University College of NSW, Australian Defence Force Academy; Dr Jim Burgess, Antarctic Geographer of University College of NSW; Dr Robert Headland and Philippa Hogg of the Scott Polar Research Institute, Cambridge University; and Dr Cathy Mead of the Commonwealth Department of Health. I was also assisted by several research grants from University College of NSW for which I am grateful. Rick Hosking introduced me to the Flinders Ranges and shared his extensive knowledge of the area with me.

Several friends and colleagues read the manuscript or parts of the manuscript in its various stages of composition. My sister, Karen Bick, offered her encouragement and advice in the early stages, and Dr Catherine Pratt gave generous help and critical commentary on various parts of the manuscript. Victor Kelleher read and commented upon early versions of the Scott material. It was also at his insistence that I wrote about Mawson at all. Dr Peter Looker and Professor Bruce Bennett provided support and encouragement throughout. In the later development of the manuscript, I was helped by the expert and professional advice of Jane Arms. Nikki Christer proved to be the ideal publisher and Virginia Lloyd provided the final editorial polish.

I would like to thank my daughter, Ellen, for her enthusiasm for this project and her cheerful acceptance of the disadvantages of living with an author. Lastly, my wife, Claire, has not only painstakingly read and commented upon every draft of the book, but has also lived with my Antarctic obsession with cheerful fortitude over several years—thank you. The mistakes and shortcomings are all mine.

prologue

THE SEA BECOMES ICE; the ice, sea. I'm standing on a beach of black lava at Cape Evans, Antarctica. It is summer, the day fine with a huge blue sky decorated by fragile strands of stratus clouds. Behind me are the waters of McMurdo Sound, and the ship that has brought me to the end of one journey and the start of another. That it is the friendliest season here, and that I am clothed in the latest synthetic materials designed for heat-retention, does not preclude me from registering the cold. It is 20°F. The air is sharp. As I walk up the beach towards the object of my pilgrimage, I'm breathing glass. But I gasp forward. My boots sound too loud, like an intrusion as I walk towards the winter quarters of Scott's 1911–13 *Terra Nova* expedition. I remember reading somewhere that conversations can be heard miles apart on the great ice barrier, the wind blowing the sounds through the vast vacuity. But here the scene is far from empty. Behind the hut rises Weather Vane Hill; its black volcanic rocks peeping from beneath the ice give an impression of untidiness, punctuating the contrast between the black beach and the dazzling white that stretches away to the south. On the top of the hill, outlined against the sky, is the huge cross made of Australian jarrah, which commemorates the death of Scott and his companions and is inscribed with Tennyson's line from 'Ulysses': *To strive, to seek, to find, and not to yield.* I am struck by a sense of romantic awe.

This is where obsession leads. I stand outside the hut nerving myself to enter the past. A strange interior surrounded by ice. Glad my hands are gloved, I press the latch. Inside, it's as if I've been here before. Light from two small windows leads the eye to the galley—once the warm heart of this polar home. The shelves are still full, waiting for the expected return of men whose laughter is long buried beneath the cairns of Antarctica or the mud of Flanders. Bottles of ketchup, Worcestershire sauce, Gentleman's Relish stand above a row of enamel cups. Boxes of Tate and Lyle sugar cubes, Huntley and Palmer's biscuits, Fry's cocoa and milk chocolate strike home with all the odd poignancy of art deco advertisements. Other shelves are lined with canned sardines, cabbage, scotch kale and processed veal. A battered iron kettle stands lidless on the range, scales and cooking pots haphazard nearby. All of it speaks of comforts created in extremity—the domestic possibilities of men, removed from the world of women, producing cosy barriers between themselves and a frigid world of deathly cold and darkness.

The long officers' table is still there too, and the bunks. One has a reindeer sleeping bag unfolded upon it. Another boasts a blue collarless shirt. A harness and other trappings hang over a post, suggestive of journeys still to be undertaken. These are the tenements, home of the English expeditioners from military backgrounds. Across the way are the berths of the Australian scientists, Griffith-Taylor and Frank Debenham. With them was grouped the young Norwegian ski champion, Gran, whose place on the British expedition was to be made so awkward by Amundsen's challenge for the Pole. Between the tenements and this more eccentric enclave, the jokes and arguments raged late into the polar night. The room is like a theatre—all the props in place, only the actors missing.

Some of the details, though, are not quite right. Others have been here between then and now, tampering with the past. The wall of boxes that once separated wardroom from mess deck, the class barrier, has been removed. And Scott's den has been methodically emptied of photos, pipes, clothes; as if to remember him needs no other memorabilia than his own words, or the statue in Waterloo Place. But nothing can silence the babble of voices that cuts the dustless air in that place. On the one side the precise jocularities of Cherry-Garrard joshing Titus Oates about being in love with his ponies—Etonian horseplay—and on the other coarser deeper accents of Lancashire, Northumbria and Wales mingle in blasphemous display.

In contrast to this is the end of the hut furthest from the mess deck. For beyond the table and Scott's den, against the far wall, are Ponting's dark room where so many images of the expedition first came to light, and the laboratories of the scientists. Here the atmosphere is serious. And as I stare at the dusty apparatus, the test tubes and flasks of a comic strip chemistry lab, I can feel the high purpose of the men who spent their hours here, endeavouring to understand where it was they had come to, and why.

Scott emerges from his den. It's just before dinner. His square face is thoughtful but not disapproving. He calls for hush from the jocular and attention from the scientists, before announcing plans and orders for the next day. I leave them to their meal, their quarrels and harmony.

Pulling the monkey's fist on the inside latch, I step from the fug of tobacco into the brittle air. I am not disappointed. The hut remains a bastion; it speaks of community and survival in this least hospitable of landscapes. It allows me in to the mythology of Scott. It makes me feel I have a beginning.

• • •

A thousand miles to the north-west I am travelling in a rubber dinghy over choppy seas towards another beach and another hut. We manoeuvre through a surprising archipelago of islets, known as the MacKellar Islands, before entering the calm waters of Commonwealth Bay. I have seen, on this short journey from the ship, under a bruised sky, the livid grandeur of the great ice cliffs that stretch away on either side of Cape Denison. The shore is only slightly more hospitable. It is a study in dark grey and white in which tumbled rocks and ice make harsh alternations. Mawson's hut stands on the floor of a valley a mile and a half wide, with the slopes rising to about two hundred feet on either side. On the top of the western rise, a single perpendicular like a flag pole can be made out against the sky; it is the wreckage of the cross erected to commemorate the deaths of Ninnis and Mertz, Mawson's companions on his terrible journey. The cross-tree of the sacrificial symbol is missing.

Walking up the rocky foreshore towards the hut, it is ferociously cold, and in defence I hold the flaps of my lined helmet under my chin. I encounter a seal carcass— a grisly remnant of Mawson's expedition, and a reminder of the antiseptic qualities of the air that is scouring my lungs. As I get nearer the hut, physical discomfort is forgotten as excitement takes over. Again I feel I'm walking into history. But this hut will not allow entry. It is partially derelict. Through successive winters it has been buried to its eaves in drift blown by the savage gales that characterise this part of Antarctica. The pressure has caused roof-windows, weatherboard and cladding to let the weather enter, a last unbidden guest. It is the home of ice and snow, the home, as Mawson said, of the blizzard. Even now, in summer, it is impossible to enter, the door jammed by detritus. Peering through broken skylights, no cosy reminders of camaraderie or youthful horseplay decorate the desolation. It reminds me of the rotting shacks that one occasionally sees from the road in country Australia, a poignant, decaying sign of aspiration miles from anywhere.

Certainly there have been attempts to restore the hut, and there are accounts of being able to enter via the roof, and see something of the inside. But now this proves impossible. The entry via the roof is boarded up, the hut keeps its secrets, and is as chilling and mysterious in its lifeless being as the seal carcass. All I have to ponder are some further remains outside the hut: protruding through the snow are dog chains, an old boot, some penguin skins and a stove plate—the leavings of men only too glad to get away, and thinking that no one would return to view these desolate remains.

I turn away and stare inland beyond the valley and see a gradual snow slope which leads towards the great ice plateau. Douglas Mawson stumbled from this direction after his ordeal and saw, miraculously, human figures waving and running towards him. He was the sole survivor of the three-man trek, a pale emaciated figure reaching out to embrace another long polar night before returning home. A night of tortured dreams, haunted by the ghosts of his companions, their names lost to glaciers. A night which would be interrupted by the paranoid ravings of the wireless operator who, driven to insanity, tapped out his delusions to the glittering sky.

Unlike Scott's hut, this is the place of survivors, yet it is replete with a sense of loss. It does not feel like a monument to endurance, to the triumph of reason, much less a national icon. The mind yearns to enter the mysteries of this interior, as its first inhabitants were inspired by the hinterland of the blank continent. It is the resistance that proves the challenge. Mawson, whose face until recently stared at us from the Australian hundred dollar bill, is as mysterious as this impenetrable edifice. One last time I try the door before walking away into the rising wind that whips snowflakes into my face.

• • •

And so I travel from Antarctica to Canberra to London to trace another beginning. On 27 January 1910 a meeting took place between Scott and Mawson, that formidable sledge traveller and explorer, who had recently arrived in the capital. On Shackleton's Antarctic expedition of 1907–1909, Mawson was one of a party of three who had man-hauled 1260 miles from Cape Royds to the South Magnetic Pole and back. He had also been one of the group who first ascended the active volcano Mount Erebus. In journeying to London he was looking to further his Antarctic adventures; he had a proposal to put to Scott about exploring the coast west of Cape Adare, that part of Antarctica closest to Australia.

Scott, meanwhile, had gleefully received Mawson's telegram suggesting a meeting about Antarctic affairs. He assumed that Mawson wanted to join his expedition to the South Pole. The prospect of this sturdy, young Australian volunteering gladdened Scott's heart. At six foot two inches and twenty-eight years old, both a proven sledge-traveller and scientist, Mawson was an ideal candidate for the long haul south. And there was the further attraction that Mawson would be giving up his former allegiance to Shackleton. Anything that might be detrimental to Shackleton was a good as far as Scott was concerned, for there was a history of bad blood between them.

Scott, shorter and more squarely built than Mawson, bustles out of his office into the vestibule to meet his Australian guest. He translates a slight nervousness into charm with his smile of welcome and the warmth of his famously intense blue eyes. Mawson has to stoop slightly to shake hands. Scott's welcome is not only to London, but also to his expedition. Mawson has to explain that that is not exactly what he wants. Scott is taken aback. The Australian explains his interest in exploring the coastline closest to Australia, and suggests that Scott could take himself and perhaps two other Australians and land them at Cape Adare to make a sledge trip along the coast. Scott is unwilling to commit himself to this plan and urges the importance of his own polar ambitions. Mawson's repeated emphasis on Australia and Australians has irritated Scott, who urges Mawson to consider the importance to the British Empire of being first at the Pole; it is crucial, he suggests, for England and the Empire to beat the Americans, Germans and Japanese. But Mawson is not swayed.

There is stalemate. Scott invites Mawson to join him for lunch in a small, not particularly smart restaraunt in St James's. Scott can't afford luxuries. The food is indifferent, the wine even more so. Mawson argues that to explore the coastline as he plans is of more economic and scientific importance to Australia than Pole-hunting. Scott counters that it would mean more to Australia for one of their people to be among the party first at the Pole than any achievement that could be gained along the coast. Mawson is as adamant as the rocks he studies. Scott is in turns charming, cajoling, irritated, determined. After three hours, there is no agreement. But Scott is pressing. He promises Mawson that if he joins the expedition he will definitely be one of the party to go all the way to the Pole. Scott, smoking furiously, is so relentless in his entreaties that Mawson agrees to be provisionally signed up and to give a definite answer within three weeks. They also agree to meet again the following day and talk to Edward Wilson, Scott's trusted friend and the chief scientist for the expedition.

When the waiter clears their plates he notices that

Scott has touched little of his food. Scott's large-boned guest with the tweed suit and flat cap, however, has cleared his plate. In pursuit of his goals, Scott could forget his bodily needs; Mawson was not so metaphysical.

The meeting with Wilson is not a success. If anything it only serves to make Mawson more determined to go his own way. He takes an instant dislike to the chief scientist and tells Scott that the only position that he, Mawson, would consider is that already filled by Wilson. Scott's reiterated appeals to Empire and the motherland leave Mawson unmoved. But Scott still hasn't finished his persuasions. He invites Mawson to dinner, hoping that Kathleen, his forceful wife, might succeed where he and Wilson have failed.

Although Mawson was impressed by the forthright blandishments of the vivacious Kathleen, he nevertheless was not going to be seduced into accepting a situation that ignored his own ambitions to explore the coast of Antarctica closest to Australia. The conversation that evening concluded with Scott saying that he would leave a place open for Mawson on the expedition until the ship left Australia. Mawson warned that he would not change his mind, but agreed to Scott's conditions. And so they parted, shaking hands on the steps of Scott's home in Buckingham Palace Road. Although there would be future correspondence between them, and they would be present in each other's thoughts, and although they would both simultaneously be in the Antarctic, the romantic and the scientist would never meet again. But their respective lives and legends would continue an uneasy relationship across years and continents.

PART ONE

an imperial romance: Scott's Story

Scott, Robert Falcon 1868–1912

Antarctic explorer, born near Devonport, Devon. He joined the navy in 1881 and commanded the National Antarctic expedition 1900–04 (the *Discovery* expedition) which explored the Ross Sea area and discovered King Edward VII land. In 1910 he led a second expedition to the South Pole which he reached in January 1912, only to find that the Norwegian Expedition under Amundsen had beaten them by a month. All members of his party died on the return journey from the Pole, their bodies and diaries being found by a search party eight months later.

What follows is an account of Scott's last days.

Friday 16 March or Saturday 17 March, 1912

SCOTT RESTED ON THE SLEDGE, shivering as the last heat generated from the march dissipated into the glittering air. 'March' was the euphemism he still used to describe the agonising lift and fall of the feet into the soft crust of the barrier surface that no one had predicted would be so difficult at this time of year. The sledge, a monstrous burden, played sadistically with back, shoulder and stomach muscles, barely moving with each step. They had been literally inching forward for days. Scott imagined both sledge and landscape as those forces or fates that had opposed him for as long as he could remember; he impelled himself forward by a deep, inchoate anger that left him breathless and dizzy with fatigue.

A break from this battle at least gave rest to limbs, but his mind was flogged by monotony. At this lunch, like so many before, all that could be seen from horizon to horizon was the featureless, staring white of the ice barrier that lately made him dream of English country lanes, just as the hard tack biscuit by his side made his mind involuntarily conjure hot scones and jam served in some Devonshire cottage. Such imaginations were both consolation and torture.

Scott dragged himself back to the present. He lifted the biscuit to his mouth and nibbled the slightest crumb from the corner; each morsel had to be savoured, and it would be some time before a mug of weak tea would be produced to wash it down and provide the only warmth felt at all these days. Wilson and Bowers were busy with the primus stove, trying to conserve the little oil and spirit they had left while melting enough snow to make three mugs of tea. They were all so weak and cold that even the simplest operation took an inordinate amount of time and tested the patience of them all. And, of course, the colder it was the longer it took to melt the snow and the more fuel was consumed. Their collective lives on this return journey from the South Pole were a constant calculation of fuel, rations and distance travelled between depots. The further they travelled the more brutal the equation became.

Now, however, Scott had a determined purpose beyond the mathematics of survival. He had to write his diary. Shortage of heat and light in the tent at night made this duty even less palatable then. And he had not made an entry since Wednesday. The intervening time had encapsulated such a mental and physical blizzard that Scott had lost track of the days; they had whirled like the landscape into one seething mass of drift through which no horizons could be seen, and objects only a few feet away were completely obscured. Sky and land became one, so that only within the confines of the tent could any spatial reality be reasserted. But there a claustrophobic nightmare had prevailed. The lines between right and wrong, illusion and reality, hero and villain had become impossible to define, and were only resolved in the end by the finality of action: Oates had walked out into the storm.

On 2 March Oates had revealed his blackened, frostbitten toes, but had staggered on gradually deteriorating and inevitably slowing the whole party down. Scott remembered the furious frustration that he had felt, and how he and his companions had battled to remain civil; how they hid their secret longing for Oates to die so that they could be freed from the burden of nursing him and progress more quickly. And they all wanted to be rid of the spectre of what they might become.

Oates had struggled on, day after day, until both his hands and his feet were frostbitten. It was as if the cold had penetrated him completely, gradually destroying his flesh and his spirit. The sight and smell of him were difficult to bear. Oates's face was a mass of scabs and pustulant sores; his fingers were black with frostbite; and gangrene had set into the frostbitten feet, filling the tent each night with the intense stink of corruption. Scott had followed this deterioration in his diary, but he had no heart to describe the physical horror. Instead he dwelt on Oates's courage but couldn't help noting that the dying man they all knew as 'Soldier' or 'Titus' was a terrible hindrance.

Now Scott had to tell the final story. On 10 March,

Oates asked Bill Wilson, in front of the others, if he had any chance of survival. Wilson said he didn't know. The others remained silent. Scott wondered if the truth would have been kinder; he knew that Oates was done. But they couldn't bring themselves to admit this awful reality to their companion. The following day there was another agonising conversation in which Oates asked for advice. Oates ought to have *known* what needed to be done; it was not for them to prompt him, Scott thought. They urged him to march as long as he could. Scott, after some argument, ordered Wilson to give each man thirty opium tabloids, so that all of them had the means to do the right thing if it were necessary.

But Wilson's insistence that opium was not the honourable way to die, and his Christian belief that to take one's own life was sinful, had its effect on Oates, who stumbled on in delirious agony for another four days in which they made only nineteen miles. At lunchtime on the last day he begged to be left in his sleeping bag to die, but the others forced him onwards. Scott couldn't leave him. The emotional burden of such an action was too great to be borne. And what if they survived to face the interrogation of a bereaved family and a public happy for scandal? No, Oates had to find his own way out of the dilemma; he should not ask others to bear the responsibility.

Scott thought of Oates's last night in the tent. Of how, before an exhausted unconsciousness fell upon them all, Oates had said that he hoped for all their sakes he would die in the hours that followed. There was nothing any of them could say. This appeared to infuriate Oates, and his voice took on a hoarse vehemence as he demanded why they didn't say something. A silence followed before he asked, in a tremulous quaver, that if they got through they should remember him to his mother and commend him to his regiment. Scott, Wilson and Bowers all pledged to do as he wished. Wilson muttered a short prayer that Oates would be relieved of his suffering and be taken to the bosom of his maker. As human voices fell to a terrible silence in the tent, they heard the gale howling outside, and before they slept wondered whose turn next it would be to face the darkness.

Scott had spent a night disturbed by dreams and waking dreams, his mind drifting in and out of consciousness, projecting fragments of desire and despair. He would be sitting down to a huge feast under chandeliers at a table replete with delicacies, his plate full, but just as he raised a forkful of succulent duck to his mouth a servant would snatch both fork and plate away, saying he had been too long. Then he was back on his first expedition to Antarctica, trekking with Wilson and Shackleton, hauling the last dying dog on their sledge. There were tears in his eyes as he handed Wilson the pistol with which to dispatch the creature. He could not witness the execution, and felt ashamed and humiliated by this squeamishness. Kathleen entered and told him to buck up and get a move on, that the only thing that mattered in life was joy and dancing. She tried against his will to make him dance on the barrier ice, but he kept falling down, and every time he did so, she laughed dementedly and glided on.

Through such shifts and juxtapositions the night had passed, and he had been aware on waking of hoping that Oates was gone. But something in the soldier would not readily give in. Oates woke again, and the agonising difficulty about what was to be done embarrassed them all. He was in a terrible state; his nose and eyes had bled in the night, leaving his cracked lips and cheeks stained with brown tears. In despair, intimidated by the oppressive silence of his colleagues, Oates took matters into his own hands, and crawled out of the tent. For Scott the following half hour was appalling. The three survivors said little, but went about their camp routine, two remaining in their bags while the third, Bowers, lit the stove and produced a meagre breakfast.

There was a time of awful suspense as they half expected Oates to re-enter the tent, each of them nursing the inarticulate hope that their burden had been lightened, and that now they just might win through.

After half an hour Scott could not bear the waiting any longer. He left the tent only to be confronted by blizzard and drift. He could see nothing, and he knew that to walk even as far as ten yards was to risk disorientation; he was not going to be lost in the snow himself. He returned to his companions and comforted himself with the thought that at least he had supplied 'Soldier' with a relatively painless way out; for surely in the end Oates must have swallowed the opium. At the last, Wilson encouraged Bowers and Scott to join him in singing 'O God Our Help in Ages Past'.

Now Scott clamped his aching fingers around the pencil and, pressing hard so that the writing would not shake too much, began to inscribe Oates's demise into his diary. After every few sentences he had to put back on the large fur outer mitts to prevent his fingers getting frostbitten. Then he'd brace himself again and set himself to composition, with only his woollen half mitts on. The narration had to be brief and to the point, with no emotional indulgence; Scott wished to dwell on Oates's heroism. Slowly he ground out the message:

> *Should this be found I want these facts recorded. Oates' last thoughts were of his Mother, but immediately before he took pride in thinking that his regiment would be pleased with the bold way in which he met his death. We can testify to his bravery. He has borne intense suffering for weeks without complaint, and to the last was able and willing to discuss outside subjects. He did not—would not—give up hope till the very end. He was a brave soul. This was the end. He slept through the night before last, hoping not to wake; but he woke in the morning—yesterday. It was blowing a blizzard. He said, 'I am just going outside and may be some time.' He went out into the blizzard and we have not seen him since.*
>
> *I take this opportunity of saying that we have stuck to our sick companions to the last. In case of [sic] Edgar Evans, when absolutely out of food and he lay insensible, the safety of the remainder seemed to demand his abandonment, but Providence mercifully removed him at this critical moment. He died a natural death, and we did not leave him till two hours after his death. We knew that poor Oates was walking to his death, but though we tried to dissuade him, we knew it was the act of a brave man and an English gentleman.*

We all hope to meet the end with a similar spirit, and assuredly the end is not far.

• • •

When Scott wrote this diary entry, he and his companions had been marching for four months, two weeks and three days. Their route had taken them from the hut on Ross Island, across the Ross Ice Barrier, up the Beardmore Glacier, on to and across the Polar Plateau to the South Pole. They were now about twenty-five miles from One Ton depot where, as its name implies, there was a considerable cache of supplies. One Ton depot was one hundred and thirty geographical miles from Hut Point and safety.

At the beginning of the outward journey, they had been helped to drag their sledges by ponies and dogs, but this had only lasted until 10 December when the last of the ponies was slaughtered and the supporting dog teams turned back. Since then they had been man-hauling in teams of four, and then five. For fourteen weeks now they had hauled a sledge sometimes weighing as much as 800 pounds and rarely less than 500 a distance of approximately 1475 miles.

By modern nutritional standards, their diet was deficient in both calories and vitamins. Since starting back from the Pole on 18 January, their supplies had been failing further because of the increased time taken between each depot. The climatic conditions and the various surfaces encountered were also formidable. In November and December, for example, crossing the barrier for the first time, the surface was often soft, with the ponies sometimes sinking up to their knees and the men struggling along beside them.

The temperatures steadily increased. On 20 November the temperature at night was $-14°F$ and rose to $+4°$ during the day. But by 5–6 December snow was melting as it fell, the temperature being $+31°-33°$, and the men were laid up in their wet sleeping bags because of blizzards.

Conditions on the Beardmore Glacier were even worse. On the Polar Plateau, although some of the going was good, the conditions deteriorated as they neared the Pole and Scott records 'heavy' and 'stiff' pulling. The snow waves, or sastrugi, which varied in height from eight inches to five feet and over which they had to haul the sledge, were another burden. And there was also the nervous strain of dealing with crevasses, which could hurtle men, animals and stores to oblivion if they were not carefully negotiated. As the return journey began, the temperatures began to drop, and by March, with autumn setting in, they are difficult to imagine. At the lunch break, when Scott wrote his diary entry about Oates, the temperature was an inconceivable $-40°F$.

Another problem was altitude, which not only made the pulling harder, but also increased dehydration. On the outward journey, they began at sea level, and at the foot of the Beardmore Glacier they were at 170 feet. At the Upper Glacier depot they were at 7151 feet, and from there the Polar Plateau rises to 9862 feet one and a half degrees of latitude from the Pole. The Pole is at 9072 feet.

These stark figures can only suggest the longevity and magnitude of the physical effort and suffering entailed.

It is at such extremes that language staggers and fails. It is enough to say that for the four and a half months of the journey the men rarely had a comfortable day and night. When the weather was warm, they hauled in their singlets, but this exposed the skin to sun and wind. Their lips were chapped, they were wind- and sunburnt. At night their clothes, wet with sweat, first froze and then, in their sleeping bags, melted, leaving them lying in wetness—all this less than two months into the journey. After four and a half months they were all suffering from malnutrition—scurvy, berri-berri, pellagra, frostbite and, in Scott's case, chronic indigestion. They were perpetually thirsty and cold. So why did they do this to themselves? What manner of men were they? What psychology led them to this extraordinary self-flagellation? What hatred of the body motivated them to punish it so much?

• • •

A sudden irregular jolt on the harness brought Scott up short. Wilson had stumbled going forward, and Bowers sank to his knees. Wilson struggled to his feet again and said, 'I'm nearly done.' Bowers agreed. They were all cold. 'Just a little further,' urged Scott. But as they heaved again, he could feel his companions were not willing, and he too felt breathless and exhausted. After a few more paces he cried, 'Spell-oh!' and they decided to camp.

This did not mean immediate rest. Pitching the tent was an arduous business, particularly in a strongly gusting wind. First of all, snow blocks had to be cut from the barrier surface in order to weight the thirty-inch valance at the bottom of the tent. It was the turn of Scott and Bowers to perform this excruciating task while Wilson unpacked the tent and tent-poles.

On his knees in the icy gale, hacking into the snow with a pick, Scott's spirits slumped, and it was all he could do not to let misery overwhelm him. The effort was almost too great. Breathless, dizzy and freezing cold, he felt the impulse to weep. But he continued his exertions as Wilson, his task completed, came over to help. It took the three of them the best part of an hour to cut out the twelve blocks that were necessary to hold the canvas down in such a high wind.

Then, with freezing fingers, the six long bamboo poles had to be slotted into the heavy canvas pole-cap. This was an operation impossible to perform without removing the outer mitts. With trembling fingers, risking frostbite, the poles had to be quickly locked into position. They took two each, and soon the skeletal frame was ready to receive its outer garment.

But this was not easy either. The double skin of the tent had to be placed over the poles and then weighted down with the ice blocks before it blew off again. This was difficult enough to do with four fit men. With three in a state of near exhaustion it was an exercise in refined torment. Scott and Wilson, being the tallest, held the tent to windward and, as the gale billowed into the tent, hauled it over the frame. Bowers then raced around to try to get the ice blocks in place before another huge gust lifted the whole billowing mass of canvas up and off the frame. Then the frame blew down. They had to start again.

By the time they had succeeded in getting the tent up, they were all very cold and nearly beyond further effort. But after a while the idea of some food and a little to drink inspired them, so that slowly they began the routine with ice and cooker which led to some meagre sustenance.

Afterwards they huddled into their sleeping bags to drift and doze, hoping that sleep would nurse them soon and without nightmare.

Scott's thoughts were of Oates. He was a man, Scott reflected, who epitomised so much of the English upper-classes. Oates's exaggerated carelessness and eccentricity in matters of dress, his reserve, his unwillingness to enthuse, his contempt for outward show, all spoke of a man with the confidence of high birthright and wealth. Scott had often felt that Oates secretly despised him, but he could hardly tell. Certainly they had had their rows over the ponies, none more so than when on the depot journey Oates had urged him to travel further than One Ton depot and cache the carcasses of pony meat. If he had listened to Oates's advice, they might even now be basking in the reassurance of renewed food and fuel supplies.

Such thoughts were uncomfortable. Easy to be right in hindsight, Scott thought. Not so easy to make decisions at the time. And it wouldn't have done to let Oates think he could rule the roost. Scott had been painfully aware of Oates's superior class and educational background and knew that he had to assert the authority of his military rank so as not to compound this disadvantage. At least I've given him a decent epitaph, thought Scott. And as he neared sleep, incongruously he thought of the night in the hut when Oates had given his lecture on horses and ponies. His laconic delivery had brought the house down. At the end, Oates had told a story about a dinner party at which he had been a reluctant guest. One of the young women of the party had been late, so the others had started dinner without her. When she arrived flustered and embarrassed, she attempted to explain. 'I'm sorry but that horse was the limit,' she said. 'Perhaps he was a jibber,' said the hostess, coming to her aid. 'No,' the young woman replied wide-eyed, 'he was a fucker. I heard the cabby say so several times.' Only an aristocrat could get away with a story like that. And yet, thought Scott, it was the aristocrat and the working-class Edgar Evans who had failed first. Perhaps there was something to be said for middle-class tenacity and ambition after all.

• • •

Oates belonged to Gestingthorpe Manor, which had been in his family since the time of the Domesday Book. He was educated at Eton before joining a cavalry regiment. His predilections were for those horse sports so beloved of his class—polo and the hunt. He was in India with his regiment before volunteering for Scott's expedition; he paid a thousand pounds for the privilege and was accepted by Scott sight unseen. Writing to his mother, his only surviving relative, about his decision to volunteer for the Antarctic, he says with sardonic understatement that the climate in polar regions is 'very healthy, though inclined to be cold'.

Given his experience with horses, Scott put Oates in

charge of the ponies but not, unfortunately, of their purchase. Oates recognised as soon as he set eyes upon them in New Zealand that they were a poor lot, and in trying to get the best out of them he was led into several wrangles with Scott. During the winter, Oates would repair with his friend 'Dearie' Meares to the stables to be near the ponies and to sit and yarn, pipes clenched between teeth, comforted by the warm animal smell and the sweet resonances of tobacco. In this atmosphere of intimacy, they could confide their disappointments with Scott to each other: what a fussy devil he was, a hopeless deskwallah; how he'd rather consider the sartorial effect of a pair of fox's puttees than come out and look at the ponies' feet; how he had a face like an old sea-boot when things were not going right.

On returning to the hut, Oates would retreat to the area known as the tenements, which he shared with four companions: Meares, Atkinson, Bowers, and Cherry-Garrard. Their motto was 'down with science, sentiment and the fair sex'. Opposite them were the scientists, Griffith-Taylor and Debenham, both Australians, and the Norwegian ski expert Gran. Between these groups there was much banter and rivalry on the political issues of the day, particularly the question of female suffrage, which was vociferously supported by Debenham and Griffith-Taylor (the latter was known in the hut as 'Keir Hardie'), and equally staunchly opposed by Oates and the others.

Oates was identified as a misogynist early in the expedition. In the wardroom of the *Terra Nova* en route to Antarctica, a chorus of voices was often raised, taunting 'Titus' or the 'Soldier' to the tune of 'Cock Robin':

Who doesn't like women?
I, said Captain Oates,
I prefer goats.

The only woman Oates liked was his mother. Otherwise he preferred to see men play the role of women in his life. He referred to his particular friend, Atkinson, as 'Jane' and Meares was his 'Dearie'. In the wardroom Atkinson and Oates were known as 'Max and Climax', and at least one of the cabins was named 'The Abode of Love'.

But physical contact between the men, in public at least, was more about pain than pleasure. Both on board ship and later at winter quarters the mess was very boisterous, and a favourite game was called Furl Try Gallant Sails, which consisted of a fight in which the object was to tear each other's clothes off. Oates was always in the thick of such ragging. On Christmas Eve 1911, when the ship was stuck in the pack-ice, he and 'Jane' entered the mate's cabin and demanded the return of twenty matches that they claimed had been taken from them. There was a tremendous scrap as they wrestled and rolled on the ground, ripping and grabbing at each other's clothes often with great success. Within a few moments Campbell was hanging over the ship's side being sick while, to the astonishment of a few quieter members of the party, the wardroom door burst open and Bill Wilson staggered in completely naked, dragging Oates along on his back.

There is a great deal of frustrated sexual energy about all this. 'Ragging' seems to be about the displacement of desire into sado-masochistic competition. The body used as an instrument of power and punishment can be tolerated naked; as an instrument of sexual pleasure it must be subdued, punished and chastened.

Oates was not only physically aggressive to some of his companions but also harboured the prejudices of chauvinism. On the return march from the depot-laying journey in February 1911, Oates said to the young Norwegian, Gran, who was known to his colleagues as 'Trigger', 'I say, Trigger, nothing personal you know, it's just that I hate all foreigners. All foreigners hate England. Germany leads the way. But you're all the same. Just waiting to destroy England and the Empire.'

Before the astonished Gran could reply, Bowers intervened. 'Could be something in what you say, Oates, but all the same I'll wager what you will that Gran would be with us if England is forced into a war through no fault of her own.'

'Would you?' asked Oates.

'Of course,' Gran said.

And from then on they became firm friends.

This is more than can be said for Oates's relations with Scott, which deteriorated further as the expedition went on. Immediately before setting out on the Polar journey, in a letter to his mother, Oates wrote that he 'disliked Scott intensely' and would chuck the whole thing in if it was not for the fact that it was a British expedition and the Norwegians had to be beaten. Oates said that Scott had always been very civil to him, and that superficially they got on well. But in Oates's opinion Scott was 'not straight' and put 'himself first and the rest nowhere': when Scott had got all he could out of you, it was shift for yourself.

In the normal course of events, it is part of Oates's code as a 'gentleman' not to make such negative feelings known to Scott, but under the pressure of dying, due to mistakes whose responsibility could easily be left at Scott's door, it is difficult to believe that this code was strong enough to be upheld.

It is something of an irony, then, that Oates's memory and reputation are embedded in English culture by means of Scott's account of his death. What the diary does *not* say, however, is what actually happened. Did Oates remember his mother and his regiment before stepping outside the tent, or before he went to sleep the previous evening? How much conversation was there that morning before Oates walked into the blizzard? And was any overt or covert pressure applied before he did so?

Instead of explanations, Scott's prose turns the incident into an example of patriotic heroism. The death of Oates, which prefigures Scott's own, is rendered as a heroic English gentlemanly ideal. Scott allows no trace of rancour or unpleasantness to intrude. There is no dwelling upon Oates's physical condition. Scott tries to make clear the connection between suffering, sacrifice and manliness by implying that Oates has sacrificed himself to give his colleagues a chance of survival. His bravery is dwelt on, as is the magnitude of his suffering. Here is an example of how to die with the right 'spirit'. Scott is anxious to assert that he and his surviving companions have acted honourably and stuck to their sick companions to the last. By implication, they too are English gentlemen.

Sunday 18 March, 1912

To-day, lunch, we are 21 miles from the depot. Ill fortune presses, but better may come. We have had more wind and drift from ahead yesterday; had to stop marching; wind NW., force 4, temp −35°. No human being could face it, and we are worn out nearly.

My right foot has gone, nearly all the toes—two days ago I was proud possessor of best feet. These are the steps of my downfall. Like an ass I mixed a small spoonful of curry powder with my melted pemmican—it gave me violent indigestion. I lay awake and in pain all night; woke and felt done on the march; foot went and I didn't know it. A very small measure of neglect and have a foot which is not pleasant to contemplate. Bowers takes first place in condition, but there is not much to choose after all. The others are still confident of getting through—or pretend to be—I don't know! We have the last half fill of oil in our primus and a very small quantity of spirit—this alone between us and thirst. The wind is fair for the moment, and that is perhaps a fact to help. The mileage would *have seemed ridiculously small on our outward journey.*

Scott felt the power of his pencil to impose order and control upon the situation. Writing, he reflected, was a lightening of the burden. In a few brief sentences the accumulation of emotion, the experience of weakness, fear, hunger and pain could all be made to seem within his command. No self-pity or gloom must intrude. Pain should only be mentioned in a matter-of-fact manner and not dwelt upon. Understatement put his foot in its proper place. And the vagaries of fortune had always to be noted. Scott grinned as he thought of his stiff upper lip literally freezing into place. He would compose himself into a model Englishman.

A certain resignation was beginning to take over. Scott remembered the impatience and irritability with which he had reacted earlier in the expedition. But why should he be impatient now? To survive with one leg and return to England having failed in his primary objective, *and* having lost two of his men, was not an enticing prospect. Scott heard the seductive whisper of martyrdom. It allowed him to write cheerfully, as if to survive or not to survive was an exciting wager.

• • •

Scott croaked 'Heave!' and the three of them lurched a step forward, feeling the jerk of the sledge, a dead weight hauling them backwards, even as they strained ahead. The wind, perversely, also impeded them. It should have been at their backs blowing from the Pole. Instead it whipped ice crystals into their faces, blowing from the north-west. It was half an hour since lunch, and Scott knew by recent experience as much as the uncertain feeling in his stomach that they would soon have to halt again. Wilson was the first to need relief. Stopping with an apology, he hurriedly fumbled with the awkward leather strapping of the sledge harness. Bowers and Scott followed suit, for they were all in similar need.

For Scott this was the ultimate misery. Even though they attempted to maintain an element of privacy by keeping their backs turned to each other, voiding the bowels on the march like this was a terrible ordeal. There was the difficulty of undoing the lampwick that secured the windproof outer garments, but also the grotesque physical gymnastics entailed by squatting while exposing as little flesh as possible to the savage cold. Even at the best of times, in privacy, Scott hated this physical function. The thought of it often made him dry retch in the performance. Now it seemed the most refined of the various tortures that he endured.

It was thus with more than one sense of relief that they harnessed up again and continued. For all the back-breaking exercise, Scott could not get warm. But the physical discomfort tugging his mind back to the pain of his body was not always strong enough to prevent memory and imagination leading him down different, though not always less hurtful, tracks.

Cursing the sledge at his back would trigger images of all that had impeded him in the past. With a mixture of anguish and bitterness, the example of his father often came to mind. How that gentle, dreamy fellow who loved his garden had let the family business go, sold it, then spent the proceeds keeping up appearances at the family home, 'Outlands'. Scott remembered his early boyhood with a pang; how the garden had been the site of his first adventures in which he and his brother Archie would rescue his sisters from marauding savages, all of them lost in some imperial outpost of the mind. But at thirteen, dreams gave way to duty when he left home to attend the navy school at Dartmouth. Years later, his father's money had run out, the family home was let, and his father had been forced to take a job as a brewery manager in Somerset.

If this slide down the subtle scales of the middle class had been precipitated by his father's feckless life, it was pushed headlong by his death. Scott and Archie had then to shoulder even more of the family's economic burden. And within six months of his father's demise, Archie, the friend of Scott's youth, was also dead. He had contracted typhoid fever in Africa where he had been serving with a West African regiment, the Hausa Force. Despite one sister going on the stage, another training as a nurse and the two youngest going into business as dressmakers, Scott was perpetually harassed by financial obligations after his brother's death. As a young man in his twenties without money to spend, he had been unable to drink with his fellows in the wardroom. His uniform had grown moth-eaten, the gold braid of his epaulettes tarnished. He had little opportunity of courtship, for he could not afford marriage, and he had not the high and aristocratic connections of so many of his contemporaries, making promotion hard to come by.

Scott remembered his father's sad, defeated face, as he

had so often done before. All his adult life, it seemed, had been a constant battle to crush those parts of himself that were like his father, and in that way avoid a similar failure. Here was a spur to work ferociously for success, to banish the soft, the gentle, the easy, and to cultivate the harder virtues of a proper man.

As he hauled the sledge another inch forward, he tried to persuade himself that at least he had done *something*. They had reached the Pole, even if they were there second, and if they survived they would have undertaken a longer journey than Amundsen, and they would have man-hauled. Man-hauling, Scott reflected, was the cleanest, the finest way of polar travel, obviating as it did the killing and eating of draught animals.

So Scott laboured to transmute the racking pain in his back, shoulder, calves and thighs into a symbol not of failure but the possibility, at least, of success. Although he now had to drag his frostbitten leg in an ungainly limp, he felt certain that the weaknesses of his youth had been conquered. He had been asthmatic and coddled by his mother as a child. From an early age he remembered the fits of lethargy and abstraction which later in his life had threatened him with a disabling sense of meaninglessness and depression.

All this had to be fought, as did other mental and physical treacheries. Sea-sickness, fainting at the sight of blood, and vertigo had all been fearfully experienced and battled. Now, Scott thought, I have nearly beaten them all. For a moment his mind filled with the heady elation that he had experienced as a thirteen-year-old cadet, whose first test of fitness for a naval career was to climb the rigging to the top of the foremast of the training ship

Britannia. He could recall the tingling in his hands and feet as he mounted the swaying ladder, the coarse, abrasive texture of the rope, the cold sweat, the dizzy fear as he ascended, all with absolute clarity. He also remembered the hoarse encouragements from the decks below, when he had stopped frozen by terror. 'Be a man!' was the imprecation, and he had somehow continued to make his unwilling limbs, step by step, mount the bidden way. But what rewards there were on feeling the deck beneath his feet again. He had almost swooned, but again commanded himself and tasted the joy of self-conquest.

Now, step by step, Scott dragged himself forward. Each step was a victory. And though the weather was so against them, with the grey sky melting into the horizon and enclosing them in a lowering world, he could feel no despair. There was pain, but no depression.

They did not get much further before they all agreed to call a halt to the torture. It took them over two hours to get the tent up, and after their exertions they just lay inside it for a time to recover. All three of them were still miserably cold and shivering. As soon as he could, Scott began to get things organised inside the tent. There was an established routine for stowing individual gear, food, and cooking utensils. Any snow that they had brought in with them was brushed off the floor cloth and out of the tent with a small brush that Scott had insisted should be carried by each sledging team. Order had to be maintained. Even though they were travellers in the wilderness, Scott believed they had to behave in a civilised fashion. He knew his two companions agreed. As during the struggle to erect the tent, Scott hardly needed to voice a command. Wilson and Bowers knew their roles.

Wilson prepared cold food. But this was followed by the ineffable comfort of half a pannikin of cocoa each, warmed over a tiny amount of the methylated spirit that was left to them. Heat and light cheered them all, as they huddled in their reindeer-fur sleeping bags. 'Cocoa will never taste as good as this when we get back,' Bowers said. Scott and Wilson agreed. A wonderful ease enveloped Scott as he heard the voices of his companions chattering about what they would and would not enjoy on their return. For his part, it was miracle enough to feel warm for the first time in days, to know that for a while at least he was out of the gale and did not have to lift his weary feet any further. He could let his mind go without fear of where it might wander. One of the great consolations to him of sledging was that it appeared to operate on the physique like multiple cold showers. He had never been so free of sex yearnings in his life. He could think of Kathleen with a detached affection; he could make her into an idea to be worshipped at a distance, without having to endure the strange discomfort of her enfolding arms.

Wilson suggested that, since it was Sunday, they should finish the day with a song and a prayer. Scott and Bowers readily agreed. Soon they were struggling to raise their voices above the flap and bang of the tent's canvas, with the haunting sentimentality of 'Abide with Me'. In the past Scott had always conformed to the rituals of Christianity but had deep reservations about its substance. Now, however, Wilson's faith was an inspiration, and Scott's feelings were elevated to a plane wherein thankfulness for the presence of his comrades was indistinguishable to him from thankfulness towards God.

Clinging to this momentary beatitude, he fell asleep.

Monday 19 March, 1912: Lunch

SCOTT SHOUTED 'HEAVE!' AGAIN, felt the strain, and thought his back might snap. The sledge did not move. Looking at the grim and strained faces of his companions, Scott suggested that they take a spell. He would use the time to write his diary while the others organised a watery brew.

Scott had always wanted to write, but felt too buttoned up, too self-conscious, too restrained. He admired those free natures like Kathleen and his friend Barrie, who were somehow able to relax and express themselves without inhibition. But now he was learning. It was, he thought, true as so many poets and writers had asserted before: in order to write you had to suffer. But you had to make of that suffering something else—simply whining would not do.

> *We camped with difficulty last night, and were dreadfully cold till after our supper of cold pemmican and biscuit and a half a pannikin of cocoa cooked over the spirit. Then, contrary to expectation, we got warm and all slept well. To-day we started in the*

usual dragging manner. Sledge dreadfully heavy. We are 15 and a half miles from the depot and ought to get there in three days. What progress! We have two days' food but barely a day's fuel. All our feet are getting bad—Wilson's best, my right foot worst, left all right. There is no chance to nurse one's feet till we can get hot food into us. Amputation is the least I can hope for now, but will the trouble spread? That is the serious question. The weather doesn't give us a chance—the wind from N. to NW. and −40° temp. to-day.

Scott felt relieved as he finished his diary entry, even though this signalled the time to don the harness again and continue the struggle. There had been many moments that morning when all three of them had pulled together and the sledge simply did not move. The surface was still terrible. It was made up of ice crystals blown by the wind into a thick crust above the solid barrier ice. Lifting one's foot from it felt as if one was weighted with ball and chain, as if the land wanted to hold you fast in its frozen embrace. It was a point of honour with Scott that he should pull as well as his two companions despite his frostbitten foot. Though at forty-three he was the oldest (Wilson being thirty-nine and Bowers twenty-eight), it had always been, and still was, a matter of considerable pride that he could show the way physically. Now, though not the fittest, he was determined to match the others in effort. He would not be left behind.

This was what leadership meant. When it came to man-hauling a sledge, moral leadership was the same as physical leadership. All through the journey Scott had striven to drive his team faster and further than the others. It was the only way. And how proudly he had been able to write his little notes to Kathleen to be taken back by the support parties. He had been able to tell her again and again that his party could easily hold its own, that he could 'keep up with the rest as well as of old', that he was exceedingly fit and could 'go on with the best of them'. He had led the business 'not nominally but actually'. He had 'lifted other people out of difficulty'. Nobody could say he wasn't fit to lead through the last lap. Kathleen could not possibly be ashamed of him, he thought.

As the three of them now inched forward, bent double against the blast, Scott tried to inspire himself with thoughts of the terrible journeys he had survived on the *Discovery* expedition of 1902–03. On the southward journey Shackleton had collapsed—a sure sign, Scott thought, of the man's moral weakness. And on the westward journey he had had to send three of his men back. Only he, and the seamen, Lashly and Evans, had stuck it out, though they came very close to disaster. But Scott had been inspired by their endurance. It made him feel strong. It showed those who talked of the degeneracy of the British race, and the working classes in particular, that men of the bulldog breed still existed. That's what all the working classes were in need of, thought Scott. Naval discipline, moral leadership—it brought out the best.

As he bent forward, he cursed his body for its weakness now. He remembered the fine words from his diary of the *Discovery* expedition extolling the virtues of man-hauling. He remembered writing of the 'fine conception which is realised when a party of men go forth to face hardships, dangers and difficulties with their own unaided efforts, and by days and weeks of hard physical labour succeed in solving some problem of the great unknown'. 'Surely,' he had asserted, 'in this case the conquest is more nobly and splendidly won.' Sledging is the way to find the 'truth' about a man; 'frauds' are quickly exposed. So it had been on this expedition. Scott thought of Gran, the young Norwegian. After only a little work the young man had complained of cramps in his legs and hobbled about pathetically. What a vain, idle fellow he'd turned out to be. Not a good advertisement for his country. Scott had upbraided and humiliated him in public to try to bring him up to the mark. But he had continued with his charade.

Then it was easy to tell the fraud. But what of now? As they strained in their harnesses to move the awful weight behind them, Scott had taken to daydreaming about being met by Cherry-Garrard and the dogs. Did this make him and his companions frauds? Surely not, Scott thought. They had done the chivalric thing. They had embarked on their quest like knights of old, relying on themselves and their strength, questing over the ice, true to their chosen chastity, to prove themselves to each other and to the women who waited at home.

Well, Scott thought ruefully, it is easy to tell such stories. To write about fine conceptions. It was difficult, as he wheezed and choked with fatigue, as he dragged his frostbitten foot behind him, as he pulled till his guts were wrenched inside out, to quite believe in it all anymore.

Tuesday 20 March, 1912

SCOTT DRIFTED BACK TO consciousness for what seemed the hundredth time. He had spent all night between sleeping and waking, due not only to discomfort, but also to the ominous boom of the tent's canvas which suggested a blizzard outside. The night before they had camped as usual after staggering a further heartbreaking four and a half miles after lunch. They were now only eleven miles from the depot. Cold pemmican and watery cocoa had been their miserable supper before forcing themselves into their frozen sleeping bags. The wind had risen during the night, making Scott anxious. A blizzard meant delay, and delay they could not afford.

Now, he perceived, it was a fraction lighter in the tent, and he wondered for a while if it was worth taking his arm out of the sleeping bag to consult his watch. He was lying curled up with his hands under his armpits and was, at last, relatively warm. But this, though a blessing of one kind, might be a curse of another. For when snow fell, the temperature rose, but the combination of wind and heavy snowfall made travel impossible. Navigation was simply too difficult under those circumstances.

It was miserable to think of moving. Only the notion

of food at One Ton depot inspired any impulse to action. But then the further equation came to mind. If they got to the depot, could they manage the extra one hundred and thirty miles back to the safety of Hut Point? As he thought of this, Scott allowed himself the momentary fantasy of arriving at One Ton to find Cherry and Meares with the dogs to help pull them back to safety on full rations. He imagined with unspeakable yearning the sensation of eating whole sticks of chocolate one after another. The way one could hold a piece on the tongue, letting its texture dissolve to velvet, bittersweet. And to be carried home, without lifting his weary feet further, was like a child's dream of flying.

But one had to be a man. What kind of a man was it who had to be carried home? And what kind of a man would be left after a surgeon's knife had dealt with his frostbitten foot or leg? Scott felt again the impulse to weep.

On this occasion, however, he was roused by Bowers' voice calling, 'Rouse out! Rouse out!' just as if they were at their happiest in the early days of summer sledging. Then nobody had needed second bidding, because it was rarely comfortable enough in the bag to encourage dawdling, and the promise of hot hoosh for breakfast encouraged activity. Now, so weakened and with so little food, the temptation to lie still was very great.

But Bowers' voice and Wilson's immediate 'Right-oh' shamed him into action, and he slowly, painfully uncurled and began to wrestle with the toggle of the sleeping bag. Wilson, on seeing this struggle begin, and aware of Scott's badly frostbitten foot, urged his leader to remain still until he and 'Birdie' had ascertained the state of the weather. But Scott's pride would not allow such indulgence, and soon he was sitting up with the others, struggling to swap day socks for night socks and to push painful feet into frozen boots.

These operations, so simple in everyday life, were grotesquely difficult in such low temperatures. They kept their socks in the inside pockets of the heavy woven jackets they referred to as 'pyjamas' in a mostly futile attempt to thaw and dry them. Now that they were wearing all their clothes all of the time, there was an ungainly wrestle to actually locate the socks. Then, having changed, the night socks had to be replaced in the same pockets. But the worst part of the operation concerned the reindeer-fur boots known as finnesko. Once having removed these the night before, each man had observed the ritual of trying to pummel the frozen boots into the appropriate shape for wear the following morning. Even if they succeeded in approximating the boot to the shape of their feet, the reindeer fur froze so hard that it was immensely difficult and time consuming to get them on. Now, weak and shivering, the struggle was terrible, and all three men cursed as they battled to get sore and frozen feet into rock-hard boots.

Scott was in a state of high emotion which, even as he experienced it, reminded him dimly of his childhood impatience—the way his stomach would knot with fury at the slightest delay to the fulfilment of his desire and he would be menaced by tears of frustration. Now he felt himself so reduced, and fought not only with foot and boot, but also to control his temper. He also felt a rising tide of panic threatening to engulf him. His foot was black to above the ankle, and ominous green tinges were beginning to show. As he attempted yet again to lift his dead foot into the finnesko, he felt defeated by the triviality of it all. That putting shoes on should come to this.

He stopped. Letting the boot drop, he leaned forward with his head on his knees, and felt the tears running warm down his cheek. It reminded him of urine trickling down the inside of his leg on the first day of school. The shame was as great.

Wilson, ever mindful of his role as 'Uncle' Bill, manoeuvred himself across the tent and gripping Scott's shoulder said, 'Come on, Tru-egg, buck up.' Scott attempted to wipe the tears away before lifting his head to smile at Wilson. No one had called him Tru-egg since midwinter dinner in the hut, when one of the men rather drunkenly had given Scott's nickname away in a toast. No one would normally dare to risk such familiarity with the 'Owner'. Wilson had gauged the moment well. Scott, with his friend's help, recommenced the struggle with feet and finnesko.

Meanwhile Bowers had his boots on and was lying prone, kicking at the entrance to the tent. This was a small tunnel of two feet in diameter, which was gathered into a bunch and tied on the inside. During the night snow had drifted against it, so that in order to get out the snow had to be kicked away from the inside. Wilson, having succeeded in getting Scott's frostbitten foot into the finnesko, left him to do the other foot himself and went to the aid of Bowers. It was exhausting work. They frequently had to stop to get their breath and rest their painfully thumping hearts.

By the weight of snow against the entrance they knew

that outside was likely to hold little promise for them. But by now they were thinking less about travelling than the necessity to 'pump ship' as soon as possible. Scott had just managed to join the others when, with a feeble cheer, Bowers broke through. After much undignified scrambling and scrabbling, they eventually crawled out of the tent.

They were met by a blizzard. The conditions could not have been worse. With heavy snow blown by a gale from the north, visibility was about five yards. They could only just see the sledge. There would be no travelling today.

Each of them moved a yard or two away from the tent and away from each other to urinate and defecate as quickly as possible. Scott's nerves and sensibilities were so battered that embarrassment or distaste were beyond him. All he wanted to do was regain the tent and his sleeping bag.

Once back inside the tent Bowers and Wilson sat on their bags, agreeing with Scott that they couldn't go on today and staring at their meagre rations. They were all chilled, but they had so little fuel that using any for a warm drink was going to be a decision which needed careful discussion and agreement. Similarly the food, enough for two days perhaps, needed to be eked out with great deliberation.

For now they agreed neither to eat nor drink so that they had something to look forward to later in the day. The struggle with finnesko accordingly started again, for the boots were too bulky to go into the sleeping bags.

Two and a half hours after waking, they were all back in their bags shivering and cursing. 'We'd better have a song,' said Bowers. Uncle Bill agreed. These two had been in tight spots together before on the winter journey to Cape Crozier when they had experienced temperatures as low as −66°F. Then as now, hymn singing was a great recourse and solace. They led off with 'Eternal Father Strong to Save'. Scott joined in with chattering teeth, and thought how lucky he was to be with these two men. He could not, he thought, be with finer friends.

• • •

Wilson, Bowers and Apsley Cherry-Garrard had survived what they referred to as the worst journey in the world during the previous Antarctic winter when they had man-hauled through the polar night to retrieve eggs from the penguin rookery at Cape Crozier. They had marched through darkness in ferociously low temperatures and lain in their frozen sleeping bags at night, iced in and shivering for hours on end. After a horrendous time they arrived at the Cape and built themselves an igloo in the ice and pitched their tent outside to protect their gear. One night inside the igloo the blubber stove spurted a blob of boiling oil into Wilson's eye. He lay all night trying to suppress the groans that involuntarily broke from him. He later confided that he thought he had lost his eye.

Worse was to come. A tremendous, storming blizzard blew their tent away, and then removed the canvas roof of the igloo. For forty-eight hours they lay in sleeping bags like frozen coffins, exposed to the blast, without food and thinking all the while that this might be the beginning of the end. Their answer was to howl out hymns as loudly as they could. After all this, they had to retrieve their tent and make their way back to winter quarters with damaged gear and clothing and inadequate food. Throughout, the men attested, there were no cross words. They remained close friends.

Yet Bowers and Wilson superficially seem the most unlikely of companions. Physically, they made an almost comical pair. Bowers was self-consciously ugly, being five feet four, weighing twelve stone five and having a round face dominated by an enormous wedge of a nose. Wilson was tall, broad, and slim with regular features that could in some dispositions be conceived of as handsome or in other moods might convey asceticism. Bowers came to the expedition from the Royal Indian Marine where he had served five years as sub-lieutenant and latterly lieutenant in command of a gunboat. His education had been on a naval training vessel and had not continued beyond his sixteenth year. Wilson had no military connections at all. He was a trained medical doctor who disliked practising medicine and worked as a naturalist. Between going with Scott on the *Discovery* expedition, and embarking on the present venture, his main occupation had been a report about the unlikely subject of grouse in health and disease for the Board of Agriculture's Grouse Disease Commission.

What Bowers and Wilson had in common, however, went beyond these superficialities. It resided in their shared Christian idealism. They were men of the kind that saints and martyrs are made of. Their impossible ideal was purity, and no scourge was too great in attempting to emulate the way of Christ. As young men, both experienced depression and nervous debility.

Wilson suffered mood swings from suicidal despair to feelings of extraordinary elation. He had such a highly nervous disposition that he used sedatives when he was faced with large social occasions. He confessed that he hated society, by which he meant the society of other people rather than the political and economic structures within which he lived. Until 1900 he was what he called a 'vicious smoker' of pipes and cigars, and waged a long battle with himself to give up the habit. He also suffered from tuberculosis. Wilson was ideal material for an Antarctic expedition where pain was guaranteed, and death a distinct possibility. It was a way of escaping society and replacing the tarnished world with 'pure' endeavour.

Like a soldier leaving for the wars, Wilson had married a month before leaving for Antarctica on the *Discovery* expedition. He and Oriana had no children and, well before his present extremity, he had written to her on more than one occasion implying that their love would be perfected in death. He was not a sensualist. Writing to his wife from winter quarters about their present separation he wrote, 'It all seems cruel and cold but it is God's will to make good stuff of us both. Anyhow you will do your duty, my brave kind lady, and your "kind sir" will do his and we will both trust in God.'

Bowers learnt his puritanical Christianity from his devout mother with whom he is said to have had a great spiritual affinity. She maintained an old-fashioned evangelical household in which there were hymns, a bible reading, and prayers before breakfast. Later, when he was away at sea, Bowers remembered family life with affection and wrote to his mother that he thanked God for the morning hymns they used to sing which were indelibly printed on his memory.

It was at sea on his third voyage, plying the route to Australia in the merchant ship *Loch Torridon* before he joined the Royal Indian Marine, that Bowers had the religious experience that was his moment of enlightenment and the gateway to a new life. There was a recurrent tension in Bowers' life between love of 'home', his mother and sisters, and his pursuit of 'adventure'. We can also guess that such feelings were exacerbated by sexual longings of one kind or another. On this particular voyage he speaks of his loneliness and the quagmire of doubts and disbeliefs into which he fell, losing all sense of purpose.

One night on deck, when things were at their blackest, Bowers believed that Christ came to him to show him why we are here and what the purpose of life really is. Christ demanded that if he chose the spiritual over the physical, a silver thread would run through all of his life. Bowers realised in a flash that nothing that happens to the body really matters.

It was essential, however, that the body remained unsoiled and was daily chastened. On the voyage to Antarctica, even in the Southern seas, Bowers had a cold shower on deck every morning. He would emerge on deck entirely naked, feeling the cold hit him like an icy wave. He felt his penis shrink and his balls tighten under the assault. But it was part of his discipline to march forward boldly with gritted teeth, demonstrating the spirit's triumph over the flesh. Arriving at the handpump he found it blocked with icicles. Undeterred, he hailed one of the sailors: 'A bucket of water, lad, as soon as you like.' The sailor raised aboard a pail of icy sea water, and made as if to put it down. 'No, no, pour it on, pour it on, and then one more,' commanded Bowers, bracing himself for the drench. The shock of the water took all thought and breath away. Then there was the stinging pain of the aftershock as Bowers scrubbed himself with a scratchy loofah, thinking all the while that soon he would be cleansed for another day. Then the second douche made him gasp, salt water rushed through his sinuses, and after another vigorous scouring he ran off down the hatch to towel himself in the cabin, his skin now a fiery red from its ordeal.

Through the winter in the hut Bowers and Wilson indulged themselves each morning with a similar ritual. They rubbed their naked bodies with frozen snow until they glowed and pain turned to pleasure. In this way they mortified the flesh and sublimated their sexuality into a relationship with Christ. No wonder they were ideal companions to die with.

Bowers, Wilson and Scott form a triangular mutual admiration society. There was nothing that Wilson would not do for Scott. Wilson wrote of him as 'a good man, a clever man, a man worth working for as a man'. Towards Bowers, Wilson expressed a similarly high regard. Bowers is 'a marvel of efficiency' and 'the most unselfish character' Wilson had ever met; he is 'a brick, irresistibly humorous, a perfect treasure'.

Bowers reciprocated these feelings entirely. He thought Wilson was the 'pre-eminent chap on the expedition ... the perfect gentleman—the most manly and the finest character in his own sex that [he] had ever had the privilege to meet'. Similarly, Bowers could not say too much of Captain Scott as a leader and as an 'extraordinarily clever and far-seeing man'. 'I am Captain Scott's man,' he declared, 'and I shall consider no sacrifice too great for the main object. I shall stick by him right through.'

Scott was closer to Bowers and Wilson than any of the other expeditioners. They are among the few who escape waspish comments from Scott's pen in his journal. Bowers is variously described as 'a perfect treasure', 'a positive wonder', a 'practical genius'; and mention is made of his 'indefatigable zeal, his unselfishness, and his inextinguishable good humour'. There are many references to his untiring energy and he is also said to be 'the hardest man amongst us'.

Wilson is no less highly commended. Scott describes him as 'the finest character I ever met ... solid and dependable ... shrewdly practical, intensely loyal, and quite unselfish'. He is said to possess a 'quiet vein of humour' and 'consummate tact'. According to Scott, there is barely a quality that Wilson does not possess. In addition to all this, Wilson is said to be a 'quick, careful and dexterous cook' who is also 'tough as steel in the traces, never wavering from start to finish'. In the latter stages of the polar journey, there is mention of his 'self-sacrificing devotion'.

Seemingly, Scott was right in his estimation of Wilson's popularity. In the journals and reminiscences of those on the expedition, he is invariably highly regarded and is thought by at least one of his former colleagues to be 'Christ-like'. But Mawson, for unexplained reasons, disliked Wilson on sight. And Scott's wife, Kathleen, described him as a 'prig'. 'I gather he thinks women aren't much use,' she wrote in her diary, 'and expect he is judging from long experience, so bear him no malice.' This was, of course, a veiled attack on Oriana Wilson also. But it suggests that, despite his marriage, Wilson was more at home with men than with women.

Wednesday 21 March, 1912

'Let's have another look?'

'There's not much hope.'

'We can't give up. There's only eleven miles to the depot.'

'Oh, we should always hope, but the question now is, what for? There's no point yearning for what may already be lost.'

'But we must try to save him and ourselves.'

'That may be impossible, whatever happens.'

The voices mingled with Scott's dream in which he had lost his way home and was wandering some unlit dank streets in a London he hardly recognised, and in which he was jostled by street urchins and ill-dressed hooligans shouting obscenities. An old crone leaning from a threatening doorway had offered a bowl of gruel which he'd snatched at before she and it disappeared into the wuthering air. And then, beneath a lamppost, outside a villainous pub, he saw his friends. Despite his tremendous thirst his friends would not enter, and they would not let him do so either. Rather they were discussing how they might save him, and as they considered his redemption, he awoke.

Bowers and Wilson were in earnest conversation. Scott, with a sick, empty feeling, wondered how he could join in. He knew he should volunteer for self-sacrifice, but the thought of being left by his friends to a slow death alone in the freezing tent filled him with terror. So the boom and flap of the tent's canvas was oddly comforting. It meant the blizzard was still blowing, and navigation still close to impossible.

In the face of this, Bowers and Wilson were talking of trying to reach the depot. Scott prepared his heroic speech. After all, he thought, it's only like writing the journal, or ordering sailors around, the assumption of an appropriately formal voice which in this case would have to be strong but quiet. 'You must go, if you can and leave me here. I'm afraid I'm nearly done, but you chaps still have a chance.'

Bowers and Wilson were surprised by the interruption, thinking that their leader was asleep. Both of them were quick to reassure Scott. 'We're talking of getting there, and getting back, sir,' Bowers said. 'There's no question of leaving you. If we can get some supplies back here, we can all still pull through.'

'That's right,' Wilson said. 'And you never know. Perhaps Cherry will be there with a dog-team to help us out. We're so close, it would be wicked to give up hope now.'

Scott decided not to press his argument, but instead alluded to the weather. 'It sounds as if it might be impossible anyway.'

'But we have to do something,' said Bowers. 'Even if we die in our tracks, it would be better than lying here and doing nothing. I'm going to look outside.'

'You must go, if you can. If you want to. You mustn't think of me,' Scott urged with as much conviction as he could muster.

Bowers and Wilson ignored this remark. The two of them began the tortuous proceedings necessary to investigate conditions outside the tent. Scott was thankful that he had made his gesture. It meant he needn't feel guilty about staying where he was. And in truth he felt little impulse to move.

...

Two hours later Bowers and Wilson were back in the tent in their bags, hoping for a change in the weather later on. They still spoke of making a move as soon as they heard the wind moderate. But for now they decided that rest was the only possible strategy.

They agreed to have a tiny amount to eat. A little pemmican on a biscuit. But they had nothing to drink. There was a drop of spirit left, and they agreed to have a warm drink later in the day. It was a pathetic repast. As the food had dwindled, on each occasion when they could wait no longer for food they ate less and less. It was as if they thought that by doing this their supply would, like the widow's cruse, never run out.

The cold, the hunger pains, the rhythmic drum of howling wind against the canvas produced in Scott a kind of delirium in which his mind obsessively returned to the scenes of conflict and disappointment in his life. He saw himself as a thirteen-year-old on the *Britannia*, a figure of shadows just before the throes of adolescence propelled him forcibly towards full self-consciousness.

There was enough pain in the recollection of that boy, struggling with mathematics and seamanship while his mind wandered to the tales of romance and adventure that were his secret passion, to convince Scott that this slender stranger in his proudly worn uniform was related to the man who lay in this awful predicament so far away from comfort. There seemed so little connection between the two, yet he recognised the threads that bind together the patchwork material of the self.

Had he not always had to fight to banish his dreaming and to leave comfort to achieve his sense of manhood in action? And was he not always haunted by an inexplicable sense of loss and emptiness no matter what he did? There was always a chasm between his imagination of how attainment would feel and the disappointment on reaching the goal. His chapped lips hurt as they stretched over his teeth in a grim leer as he thought of the apotheosis of this pattern in his life at the South Pole. It was too bitter to contemplate for long.

Now, surely, he was entitled to drift and dream and enjoy such luxury as he had never allowed himself to do before. Growing older, he had always tried to banish everything that he considered unmanly about himself, including a tendency to languid reflection and imaginative self-indulgence. Heroes of adventure were men who left the domestic sphere and all its womanly trappings to explore and conquer in the greater world. They never gave a backward glance and lived for excitement, finding their poetry in action, their passion in comradeship. Scott, with the spectre of his father's failure always before him, strove towards such manly endeavour as an antidote to the indolence with which he feared himself infected. He would not be like his father, that passive doddering fellow happy to daydream in his garden while the family fortune crumbled around him.

But self-conquest was never complete. Despite leaving the *Britannia* with first-class certificates in mathematics and navigation, despite all his striving on various ships, there always seemed to be something missing from his life. He remembered ruefully the outpourings to his diary articulating the depression he was so prone to: 'This slow sickness which holds one for weeks; how can I bear it? I write of the future, of hopes of being more worthy, but shall I ever be? Can I alone, poor weak wretch that I am, bear up against it all? How can I fight against it all ... No one will ever see these words, therefore I may freely write: What does it all mean?'

The *Discovery* expedition and his marriage had failed to answer this question. Like the inscrutable continent that he had chosen for his place of achievement, there was so much in himself unmapped, unknown, liable to trick and betray him. The conquest of the Pole Scott dimly considered no victory at all; they had been second there, had discovered nothing. The great wilderness stretched away on every side, defying his puny efforts. Perhaps the answer was death, the pristine blanket that would caress and enfold and would leave no further yearning.

His mind shifted to other moments when he had wished for comfort and found none. He remembered stumping the country for funds in 1910. He had loathed public speaking. Every night became a test of nerve invariably preceded by imaginings of how hostile the audience would be, how they would regard him as a self-serving humbug, how demeaning it was for him to be begging from public platforms. This self-laceration had been exacerbated because everywhere he went Shackleton's achievements were on people's lips, and Shackleton was an immensely charismatic public speaker; he was the kind of man who found his ego blossoming the more people there were around him. Time and again after a night spent in some draughty municipal hall, speaking to the worthies of a provincial town, Scott, comparing himself with Shackleton, had cursed himself for a clumsy, stupid and inflexible fool.

And after these occasions, because he had so little money, he had found himself staying in the kind of place that was the haunt of commercial travellers, businessmen with accents as broad as their hands, dowdy spinsters, and impecunious widows. So many rooms with their faded prints and frayed carpets had given him their damp and chilly greeting. He would light the gas mantle and close the shabby curtains, trying to create a sense of being comfortably enclosed, warm and secure from the contingencies of the world.

But none of these actions had ever had the desired effect. It was as if the night's darkness entered his being and could not be dispelled. Depression had so often settled with him in the room, like a familiar but unwelcome guest. It was in these moods that nothing seemed to make sense, all belief systems failed, and he was left with nothing. And then his sleep would be tormented by unruly dreams in which an unknown voluptuary threatened to drown him in her embrace. At first the face seemed inviting, kind. But, as she kissed him, he was seized by avid arms and legs, and her tongue filled his mouth and wouldn't let him breathe. He would awake in a slurry of shame and disgust.

These were evenings when he had tried to write to Kathleen. He would try to be cheerful in accordance with her cult of joy and happiness, but soon he'd be talking about the wretched evening he'd spent, and what a waste of words he'd expended for a mere twenty-five pounds gained—barely enough money to cover the expenses of his visit. He wrote, too, of the few blunt men who had attended his meeting in a dusty hall and who had risen to their feet in their crumpled dark suits and tarnished watch chains to question the value of attaining the South Pole. Scott had haltingly rehearsed all the arguments he knew so well, but his earnest recitation had failed to inspire them. He remembered the mocking voices interrupting his appeals to science and patriotism: 'What about money for th' unemployed?'; 'If you go t' Pole, can we come too?'; 'It'd be better grub and conditions where tha' wants to go than we've got 'ere in bloody Wolver'ampton.'

Often the letters would remain unfinished or unposted. He hated himself for all the lugubrious words written in similar moods of self-distrust and nebulous apprehension sent to Kathleen during their courtship and marriage. She seemed altogether too young, fine, vivacious and unconventional for a stolid navy man on the brink of middle-age. He thought of her tales of working with Rodin in Paris, and of her passionate, enthusiastic friendship with Isadora Duncan. He thought the dancer went too far, and said so, occasioning another brusque exchange of letters.

She had not found their courtship easy either. After all, he was mostly away at sea, and she had been tempted by a younger man, Gilbert Canaan. Three times she had threatened to call off their engagement. He could remember the phrases she'd used: 'Don't let's get married ... though much of it would be beautiful, there is much also that would be very difficult ... We're horribly different you and me ... I won't marry you ... You must stop loving a little vagabond—I've got it in my blood, dearest how I love it. The freedom and irresponsibility.' How frightened he had been of losing her, and yet it was her ability to throw off the shackles that most appealed to his imagination. Even her handwriting with its great sweeps and loops spoke of a liberation which his own small, crabbed hand could not aspire to. His only possible freedom was the white discipline of Antarctica.

All this and more had been with him in those nights and days. The world had seemed a place of chaos and confusion. The newspapers were a litany of England's troubles. Glancing from page to page he'd moved through strikes, unemployment, suffragette agitation, and the Dreadnought question. What troubled and disturbing times they were, compared with even ten years ago. The certainties with which he'd been brought up were under threat. Everyone was discontented with their place in the scheme of things. The working class were in a state of agitation, fuelled by envy and greed. Suffragettes were meddling in business that was none of their concern. The upper classes were more interested in fashion and dancing than in their obligations to the rest of society. No one *believed* in order anymore. No one believed in England.

All this had made the ice feel welcoming. An ironic bitterness filled his mind as he thought of this. So often he had dreaded this second expedition, the days and days of weary marching, the fatigue, the anxiety, the sheer cold, the haunting possibilities of failure. But after those lonely nights in cheap hotels, the purity of iceberg and snow plain, the isolation from a troubled world, had seemed seductive. A real escape for a man of action, and if the cost was suffering then so be it.

In expecting to suffer, he had not, however, anticipated his current predicament. But this was surely better than the alternative. To have stayed at home and risked no promotion, Kathleen's scorn, weary years of living death as a mediocrity, was unthinkable. Kathleen's career as an artist would be flourishing; she would stay faithful to her creed of joy, which for her was a discipline. Scott knew he could not share her belief. Such a life would have been harder than this.

Scott felt cold again and, shivering, thought of starched sheets and found himself in that desolate northern hotel of the mind again. He huddled down and wished for fire. Feeling terribly alone, he thought of the time he had shared the sleeping bag with Evans and Lashly as they fought their way up the Ferrar glacier and two hundred miles westward across the Antarctic icecap in 1903. They had neither dogs nor ponies, but man-hauled day after day. Each night they would bivouac and relish their crude rations and comparative comfort. Outside there was nothing but ice and the shrieking gale; inside was warmth and comradeship. Life was stripped to elemental simplicity, and under such circumstances bonds were formed that society made impossible elsewhere.

On that occasion Scott had lived in intimate proximity with men from a lower social class and lower rank than him, learning more of life on the mess deck than he had ever done before. He also learned to idealise these men; their unswerving devotion to him, their trust in his wisdom and abilities made him feel both needed and secure. Now, as he thought of them again, he was aware of a yearning for their presence. It was not the first time that such recollections had led Scott to crave a comfort which he dimly realised could only be given by the physical closeness of men.

Before he put this thought away from him, he allowed himself to acknowledge his regret that, as Bowers and Wilson lay beside him in their pathetic row, they did not have one of the big three-man sleeping bags they'd used on the *Discovery* expedition. But perhaps his companions would have been disgusted with such an arrangement. And in truth, relations between them were not of the easy harmonious kind that he'd enjoyed with Evans and Lashly. There was the creeping sense that he had failed as their leader, that he was responsible for their predicament. His mind returned to dwell on Evans and wonder that such a huge figure of a man should so unaccountably have failed him. At least death, he thought grimly, would be sure and faithful.

And still the tent's canvas declaimed the blizzard, and the darkness remained unpunctuated, until Birdie's voice broke through Scott's web of memories suggesting a cup of tea. The thought of tea was like being offered light after having dwelt in shadows. Scott and Wilson were enthusiastic in their assent, not bothering to consider that after this cup there would only be one more. They chose to ignore this circumstance and, as Bowers began to struggle with the primus, and Wilson shifted himself to help, they struck up a chorus of 'Hearts of Oak'.

As they croaked their patriotic defiance, Scott began to wrestle within his sleeping bag. Light from the stove meant that he could struggle with his diary while the others, who had long since abandoned writing their accounts, crouched shivering over the primus:

> Got within 11 miles of depot Monday night; had to lay up all yesterday in severe blizzard. To-day forlorn hope, Wilson and Bowers going to depot for fuel.

There was nothing else to say. Writing was painful, and the thought of the tea militated against concentration. As they worked to prepare the drink, Wilson and Bowers spoke again of making a dash for the depot. This was easy to say, but in the continuing blizzard difficult to execute. And though their feet and legs were not as bad as Scott's, gangrene was beginning to affect them all, as was scurvy and other debilitating symptoms caused by their poor diet and their continued exposure to the cold.

With the storm still raging outside, it was easy to decide to wait and see rather than to plunge into the drift, leaving Scott alone and with little chance of their own survival. Though it was unspoken, Bowers and Wilson knew there was an increasingly terrible finality in the thought of walking into the blizzard. Oates's example was only too fresh in memory.

So they clutched their tea and sang together and dreamt of limping to One Ton depot and of the oil and food to be found there.

• • •

The brevity of Scott's diary entry, the unspeakable conditions, the awful yet suppressed drama of their predicament, make it increasingly difficult to imagine or to write with any confidence about what was happening in the tent. Yet it is hard not to speculate, to feast at their extremity. Their physical condition was terrible and worsening; they were dying slowly. The temperatures were very low. What of their mental condition and the relationships among the three of them? They had been in each other's company for months under increasingly stressful conditions. But diaries and letters only speak of a cheerful fortitude despite the fact that the vitamin deficiencies they were suffering from are known to cause depression, irritability and anxiety. In its more advanced form, pellagra, which arises from a deficiency of niacin, can also cause disorientation, delusions and hallucinations.

Did their tribulations bring them closer together or drive them apart? Did they share last confidences and feelings or did their British sense of propriety and reticence command them till they died? Were there disagreements about how they should proceed? It is difficult to think of Bowers, the youngest, the physically strongest of the three, submitting to prolonged inaction easily. But perhaps he was by now too weak to care, and was determined to endure whatever his elders decided should be his fate. Scott had virtually given up hope of moving again; perhaps he entertained fantasies of rescue by Bowers and Wilson or even by other members of the expedition set out with dogs to look for them. But he knew that time was running out.

At −47°F, with a blizzard blowing, with little or no food, and debilitated by months of weary hauling, the degree of suffering must have been so intense that surely the body would have let go and allowed them at least to lapse in and out of consciousness. Exposure certainly releases the sufferer from a full consciousness of the cold. You become physically and mentally numbed, with very slow reaction times and poor co-ordination.

At this stage, and for the next few days at least, the explorers seem to have somehow avoided this condition, since both Scott and Wilson continued to write letters on or beyond 21 March. These documents, particularly Scott's, are marked by extraordinary lucidity. Is it possible that in their extremity they took small doses of the opium? This would provide a short-lived 'rush' and a feeling of well-being; reserve and inhibition might be relaxed by the drug. Perhaps this helped Scott and the others to express themselves with heroic self-delusion.

A further curiosity is provided by Bowers' assertion in his last letter that he is still strong and hopes to reach the depot. This is followed by several remarks that imply the certainty of his death. He says that the end will be peaceful—'just a falling asleep in the snow'. And that it is splendid to die with such companions. Yet there is also talk of struggling to the end.

Bowers oscillates between clinging to receding hope and the ever-increasing certainty of death. And between these positions we have his declaration that he is 'still strong'. If this was so, why is the premonition of defeat so strident? The answer may be that he was simply fooling himself, offering hope when there was none. The other, chilling possibility is that he knew he could save himself, but that he was trapped by the weakness of his companions and his obligations to them. It was one thing to reach the depot, eat plenty of food, and then continue towards Hut Point. It was quite another to reach the depot then return for Scott and nurse him and possibly Wilson all the way back. Bowers was only twenty-eight. He was the youngest and fittest, but he was also the junior officer, used to doing what he was told.

Thursday 22 March and Friday 23 March, 1912

IN THE DARKNESS OF THE tent, time began to lose its meaning. Instead there was a cycle of mental states that shifted between the blessed condition of unconsciousness, restless periods of dream, and a half-waking twilight full of longing and memory, the last rags of desire. But the worst state of all was full consciousness, when the reality of their situation and its pain impinged without respite. It was also in this state that the three of them engaged in the exhausting wrangle about what should or should not be done. Birdie continued to press the case for action before it was too late. Wilson counselled patience. Scott said little, feeling it was not his place to interfere as he was the greatest invalid. Yet, at the last debate, he had suggested that it was too great a risk to stumble into the blizzard. Their chances of survival, of finding the depot, were so slim that their best bet was to hang on in the hope of better weather. Wilson reiterated that if they did go they would certainly try to get back for him, at which Scott made a feeble protest. The conversation ended with Bowers asking the imponderable question of how much longer they could afford to wait without action. Wilson had answered by alluding to the will of God who might yet save them if the weather changed. But for now all they could do was wait, and in moments of strength attempt to write to their loved ones.

For Scott, writing was becoming an obsession. It was not merely a matter of saying a last farewell to relatives, as it was for Bowers and Wilson. Scott felt an immense emotional pressure to communicate his postscript to the story now nearly finished in his diaries. He wanted to speak in a personal way to friends, benefactors, his mother, Kathleen, and to that most unpredictable of audiences, the public. Though it was physically very difficult to write because of the cold, nevertheless in this activity Scott found his concentration absolute, and in that absorption there was both satisfaction and relief. For in writing he could explain himself and how the expedition had come to end in the deaths of five men. He could attempt to control the way people would see him. He could, perhaps, avoid disgrace.

MESSAGE TO THE PUBLIC

The causes of the disaster are not due to faulty organisation, but to misfortune in all risks which had to be undertaken.

1. The loss of pony transport in March 1911 obliged me to start later than I had intended, and obliged the limits of stuff transported to be narrowed.

2. The weather throughout the outward journey, and especially the long gale in 83°S., stopped us.

3. The soft snow in lower reaches of the glacier again reduced pace.

We fought these untoward events with a will and conquered, but it cut into our provision reserve.

Every detail of our food supplies, clothing and depots made on the interior ice-sheet and over that long stretch of 700 miles to the Pole and back, worked out to perfection. The advance party would have returned to the glacier in fine form and with surplus of food, but for the astonishing failure of the man whom we had least expected to fail. Edgar Evans was thought the strongest man of the party.

The Beardmore Glacier is not difficult in fine weather, but on our return we did not get a single completely fine day; this with a sick companion enormously increased our anxieties.

As I have said elsewhere, we got into frightfully rough ice and Edgar Evans received a concussion of the brain—he died a natural death, but left us a shaken party with the season unduly advanced.

But all the facts above enumerated were as nothing to the surprise which awaited us on the Barrier. I maintain that our arrangements for returning were quite adequate, and that no one in the world would have expected the temperatures and surfaces which we encountered at this time of the year. On the summit in lat. 85/86° we had $-20°$, $-30°$. On the Barrier in lat. 82°, 10,000 feet lower, we had $-30°$ in the day, $-47°$ at night pretty regularly, with continuous head wind during our day marches. It is clear that these circumstances come on very suddenly, and our wreck is certainly due to

> *this sudden advent of severe weather, which does not seem to have any satisfactory cause. I do not think human beings ever came through such a month as we have come through, and we should have got through in spite of the weather but for the sickening of a second companion, Captain Oates, and a shortage of fuel in our depots for which I cannot account, and finally, but for the storm which has fallen on us within 11 miles of the depot at which we hoped to secure our final supplies. Surely misfortune could scarcely have exceeded this last blow.*

Misfortune. The same misfortune that had dogged him all his life. It filled his soul with bitterness. Shackleton had not had such weather on *his* outward journey. And Amundsen could not have had such tribulations. Was he, Scott, cursed of God? Was this punishment for his religious doubt? It all seemed unaccountable. Unless, of course, the suffering and failure were willed by Providence. The image of the suffering Christ came before him, and he thought that he and his companions had taken up their cross and followed. And surely their way had been harder; their icy calvary more prolonged, their cross, that terrible sledge, more heavy. And if Christ's suffering meant love and salvation, then surely their exquisite suffering had meaning too.

He thought again of Oates and Edgar Evans, how they had let him down. Perhaps it had been a mistake to include Oates in the final party. But following the soldier's work with the ponies, Scott felt he owed Oates a place, and there was the form of it all to consider. To have a guards officer with him at the Pole somehow struck the right military note—an alliance between the services. Anyway, Scott could not have told the aristocratic Oates that he was taking Bowers instead—it would have been such an insult. So he had decided to take five men instead of four. And Scott had had no idea of the extent of 'Titus's' war wound. He'd heard rumours, of course, and sometimes seen a slight limp, but nothing that might indicate the potential severity of the problem. With a shudder Scott recalled the brief glimpse he'd had of the mass of raw pulp at the man's hip where scar tissue had dissolved, and smelt again the sweet stench of corruption from the gangrenous legs.

And then there was Evans. Scott would have staked his life on the man known to the expedition as 'Merry Taff', that giant with his endless fund of anecdotes and cheery good humour. Before his collapse he had been the ideal sledging companion. Tall, broad-shouldered, well built, he was the perfect sledgemaster and rigger, having the knack of stowing gear quickly, neatly and to the best advantage of those who had to haul. His square, handsome features had been the image of dependability. He was also intellectually brighter than most seamen Scott had encountered. Evans had a fondness for reading adventure stories and was particularly fond of Dumas, whose name he insisted on pronouncing in his rich, deep Welsh brogue as 'Dumb-ass'. His simple but amusing tricks of speech had always caused a smile among his companions. Taff habitually referred to his nose, whether red from celebration or white with frostbite, as 'my old blossom'; his skis were 'planks'.

Scott knew of his devotion. On the decks of the *Terra Nova* he had overheard someone twitting Taff, saying, 'You'd follow Scott to hell if he decided to go there', and the Welshman's straight response: 'Yes, I would as well.' Such simple sentiment had filled Scott with a rush of deep feeling. How splendid, he thought dimly, and felt he would as willingly die for Taff Evans as Evans for him: 'Greater love hath no man than this ...'

How mistaken he'd been. Scott thought with distaste of the shambling wreckage that Evans had been reduced to as they plodded through hell together. His morale and mind had failed completely. They had found him that last day, raving semi-naked in the snow. He had lagged behind for a second time on the pretext that his snow shoes had come loose. But on this occasion he had not caught up. They went back and found him dishevelled, bewildered, disoriented, his trousers round his knees.

This was the culmination of a gradual deterioration that had begun as they started back from the Pole. Evans was depressed, debilitated, his morale destroyed. Scott could not forgive him. When things were at their worst, men like Evans should have come to the fore, he thought. If England were to go to war with Germany, as his journalist friend had predicted before they left for Antarctica, she would have need of working-class men whose moral courage was beyond question, who would stick it till the end. But Evans had collapsed. Scott felt betrayed. It was difficult not to let that bitterness into his writing. He could hint at disapproval but must give the appearance of generous restraint; the world would know him as a man of principle, speaking of failure when he saw it; but also as a man of strength and compassion, refusing to rail against Evans or their fate. The writing had to be planned as carefully as the journey it described.

• • •

The failure of Evans and the deterioration of Oates are less surprising than Scott found them when viewed from different perspectives. Evans was a physically strong man with a reputation as a binge drinker. He had narrowly missed being expelled from the expedition altogether when he had got blind drunk in New Zealand and failed to negotiate the gang plank of the *Terra Nova*. Although he was fished out of the water safely, the disgrace to the expedition threatened his place in it, particularly given that he had also got drunk in Cardiff while the *Terra Nova* was in port. But he was forgiven because he was a favourite of Scott; he was taken as one of Scott's tentmates for the same reason. And whether or not it is the case that Evans's binge drinking had weakened his health and contributed to his deterioration in the Antarctic, Scott failed to consider the psychological implications of what it meant to indulge in such behaviour. The need or desire to get blind drunk suggests a need to escape the exigencies of the self and circumstances, a desire to quell excitements and insecurities in a comforting oblivion. It suggests a problem which in ordinary circumstances may be overlooked as too commonplace to matter much, but in the unforgiving extremity of Antarctica could easily become a liability. Scott might have been better off with either Lashly or Crean, who were sent back with Lieutenant Teddy Evans, and were responsible for various heroics in getting their sick officer back alive.

The immediate cause of Evans's death, however, had more to do with scurvy and other vitamin deficiencies than with a predilection for beer. Scurvy both inhibits healing and diminishes resistance to trauma and shock. Experts now believe that as a result of a trivial fall in which he bumped his head, Evans tore a small vein on the surface of the brain and a subdural haematoma—a swelling of clotted blood—began to form, which put pressure on other areas of the brain. This accounts for Evans's erratic behaviour over the last thirteen days of his life. He eventually died in a coma caused by brainstem haemorrhage, which is the typical end of such a condition if it remains untreated.

Oates was fond of a drink as well, but his collapse was physical rather than mental. He had been shot through the hip in the Boer War, and this caused a compound fracture of the joint, which sometimes caused him to walk with a limp. It seems incredible that Scott should choose someone so disadvantaged to struggle to the Pole and back. Oates's problem was exacerbated by severe scurvy causing the scar tissue around his old wound to break down. And it seems likely that Oates, aged thirty-two, was the first of the party to suffer from severe frostbite in his feet because of reduced circulation as a result of his war wound.

• • •

Scott's mind wandered over the bleak terrain of the past, trying to understand how and why he had come to this disastrous twilight in the tent. He remembered the trip to Norway with Kathleen to test the motor sledges and how Nansen, in between his flattering attentions to Kathleen, had kept repeating to him, 'Dogs and Ski', 'Ski and Dogs'. But Nansen had never been south. Dogs, ski, and furs might be all right up north in Eskimo country, but down here they were of limited value. How could you wrestle dogs up and down glaciers where skis were equally useless? Scott, like Evans, had never really got the hang of planks.

Fragments of the arguments he'd had with Meares and Oates over the dogs and ponies came back to him. Scott had overheard Meares saying to Oates in a stage whisper that Scott needed to buy a shilling book about transport. Scott had rebuked them both. And every journey they'd undertaken had surely proved him right. Man-hauling was the most reliable mode of transport—dogs and ponies were too temperamental, and the surfaces over which they had to travel too unpredictable, to ensure success. The disastrous depot journey had proved this to be true. After twenty-four days of weary struggle they'd reached the depot they named One Ton. Then Oates had argued with Scott about pressing further south. 'We should go on, sir, and kill the weaker ponies as we go and leave the meat in depots.'

'No, they've had enough already,' Scott replied. 'I won't flog them further.'

'I'm afraid you'll live to regret it, sir,' said Oates, putting a particularly derisive emphasis on the word 'sir'.

'Regret it or not, my dear Oates, I've made up my mind like a Christian. I won't punish these animals any more.'

And that had been the end of it. Oates had turned away with a grunt and a shrug. Despite everything, Scott still felt he had done the right thing. It would have been an act of criminal cruelty to drive the ponies on under the circumstances. And to shoot them deliberately was awful. He could not stand it. He was sure he was right.

If it hadn't been for circumstances they would have got through. The weather had been against them the whole way. How could he have known that they would encounter such terribly low temperatures before the onset of winter? It was inconceivable. Fate had been against them, and there was nothing he could do about it now. He had done his best. There was nothing for it but to make as dignified an ending as he could.

...

Scott's 'Message to the Public' is shaped and constructed to persuade the audience of his lack of culpability in the catastrophe it describes. Here is no scrappy, disjointed scrawl but an elegant, considered oration. There was no 'faulty organisation', every detail of food, clothing, and depots 'worked out to perfection', the party had 'fought and conquered' untoward events. What defeated them, in his mind, were a series of unforseen circumstances, 'misfortune in all risks which had to be undertaken'.

The truth was that the planning of the expedition was characterised by haste and lack of method. Scott paid some attention to motor transport, and some to ponies, dogs and skis—but not enough. He paid some attention to staffing, but much of the recruitment was haphazard and dictated by Scott's emotional involvement with some men and his bias towards others. With little knowledge of dogs or ponies, and seemingly little interest in learning, he despatched men he hardly knew to buy the draught animals for the expedition. As a result they began with poor animals. Scott refused to see or believe this. But it is quite clear, in any case, that he preferred man-hauling. He was sentimental about animals, seeing in their suffering an image of his own, a kind of anthropomorphism that led to cruelty. Just as Scott drove himself and his men to unimaginable sufferings, so he was responsible for the torture of the animals.

There were risks, too, with research and planning. Scott does not seem to have been well read in the history of polar exploration (Arctic or Antarctic), and was unwilling to follow the example of anyone not British. Esquimaux and Norwegian experience was spurned in favour of British, and specifically British naval pluck, endurance and improvisation. In the matter of food and clothing he was ill-prepared. He also had not thought through the details of his attempt at the Pole. Scott was still planning how to do the journey to the South Pole through the Antarctic winter immediately preceding their setting out. The depot-laying journey of January–March 1911 was also improvised at the last moment.

It was during this first trip that Scott 'lost' the pony transport alluded to in his message to the public, and that Oates realised the depths of Scott's ignorance. If Scott had listened to Oates, all of the polar party might have been saved. As it was they lost three ponies through exposure, exhaustion and malnutrition, and then a further three in terrifying incidents in which the ponies were cast adrift on sea-ice as Scott's party attempted to cross from the ice-barrier to Hut Point on Ross Island.

Scott's misfortune turns out to be indistinguishable from mismanagement.

The same may be said of the other 'misfortunes' to which he alludes. Of course, the weather was to some extent an imponderable—knowledge of the conditions was based entirely upon his own and Shackleton's previous experiences. Shackleton had recorded temperatures between −20°F and −40°F on the ice-barrier at that time of the year. But Scott didn't trust Shackleton and evidently had not read his rival's account carefully enough. Even so, the fact that information was scant should surely have counselled caution, particularly in the matter of provisions. But the fact that they were ultimately to rely on man-hauling meant that these had to be kept to a minimum. And so the blizzard that Scott mentions, which occurred on the outward journey at 83°, was threatening because the four-day delay meant that their rations were short.

Wilson considered the situation so grim that, with an alarming enthusiasm, he foreshadowed doom in his own diary entries. As he lay in his sleeping bag, cold, wet, and hungry, he had been reading Tennyson's *In Memoriam* and realising what a wonderful piece of faith and hope the poem constituted. It also made him feel that if he died here, as sometimes seemed likely to him, then all would be for the best. There would be no need for his wife Ory to mourn, for her trust would be like Tennyson's. Wilson's faithful fatalism precluded blaming Scott or himself for any of the disasters which followed.

The problem of weather on the return from the Pole was exacerbated by the men's physical condition. They were suffering from scurvy and other consequences of malnutrition. They were inadequately clothed. They were exhausted and dehydrated. And they were so, precisely because the arrangements of food supplies, clothing and depots had been anything but perfect.

A final straw was Scott's last-minute decision to take four men with him to the Pole. Despite the fact that all the plans for the journey south involved teams of four men who shared a tent and rations, and that the ration bags were all calculated on the basis of feeding four men at a time, Scott decided to reorganise. The night before the last supporting party turned back, he added Bowers to his own four-man team. This meant that five men shared a four-man tent and, perhaps more crucially, that Bowers and Evans had hastily to re-arrange all the supplies so that they would feed parties of five and three respectively, rather than four and four. It also meant that every time the five-man team prepared hot food, they used more oil and took more time than they would have done with four men.

This hasty re-organisation also relates directly to the 'unaccountable' shortage of oil that he found at depots on his return journey. There is no question that Scott believed that Lieutenant Evans's party had taken more than their share. In a phrase deleted from the published diaries, Scott remarked that 'generosity and thoughtfulness [had] not been abundant'. But it could be argued that the 're-organisation' left this return party short of hauling power, and who knows how accurately the food was redistributed? What is certain is that the return party was in a desperate state and could hardly be blamed if mistakes were made when measuring food and oil out at the depots.

There was also a further problem with the oil drums. It was found later that they were inadequately sealed so that leakage occurred from full drums. Here was another failure of equipment. If Scott had researched properly he would have found that Peary had encountered the same problem in the Arctic and solved it by redesigning the seals on the drums. But Scott was not one to take the advice of Americans. He was a British adventurer, a British hero. Risks had to be taken, difficulties overcome; not by planning and forethought, but by will and endurance.

This is the tale of hindsight.

• • •

Scott's writings and reveries were punctuated by sleep. He woke feeling groggy and lethargic. He had become so used to the constant noise of the tent responding to the gale outside that it no longer grated on his nerves. He was beyond such considerations. Yet through the aching hunger, the excruciating weariness, the desire for oblivion, burned the need to tell his story. His mind was haunted by the possibility that he would be blamed for the loss of life; that he would be dishonoured for only being second at the Pole. He felt as he had done when other misfortunes had threatened to wreck his career. With a shudder he recalled the moment when, in command of HMS *Albermarle* on a night exercise in the North Atlantic, his ship had collided with HMS *Commonwealth* while he was not even on the bridge. An appalling sickness of fear and shame had lasted for weeks, and though the Board of Enquiry finally exonerated him, still in the night the whispers of doubt came back to haunt him. If he had been on the bridge... if he had been more careful... if he had responded quicker to the emergency... All the possibilities of personal failure returned, and he felt his face burn with recrimination.

The same sick feeling had not been absent from this expedition either. The *Terra Nova*, fearfully overloaded on leaving New Zealand, had nearly foundered in a terrible storm before they were even close to Antarctica. And it was he who had sought the registration of the ship as a motor yacht so that it could escape Board of Trade regulations on the question of loading. It was he who had agreed to the painting out of the plimsoll line on the ship so that it could be laden with impunity. How he had cursed fate when the storm broke upon them. And how desperate those days and nights had been as they bailed and bailed, fighting to save the vessel and themselves from ignominious graves. All his life seemed to be a continual struggle with disaster, a battle to prove he could defeat the most malign of circumstances. But was this not the stuff of all heroic endeavour? Were not heroes bred from adversity?

A spasm of shivering overtook him while Bowers snored through his beaky nose and Wilson slumbered on peacefully. Scott envied them. They were followers; they did not have the pressure of justification upon them. He had to be their spokesman and his own.

He thought of the opium tabloids. They had thirty each. In such physical agony as he found himself, legs stinking and gangrenous, his skin peeling and his hair falling out, the cramping hunger pains and the invasion of that terrible coldness, surely he could afford himself some medical comfort. The brandy was long gone, most of it used to fortify Oates on his sad way. Now with the others asleep he could at least try a little to see if it helped.

A few moments later he was able to take up his pencil again. A feeling of greater clarity and wellbeing spread through him. The little lampwick taken from the finnesko he knew he would never use again burned brightly in the lid of an old tobacco tin, and somehow it made him feel hopeful and inspired. He would finish his message to the public.

> We arrived within 11 miles of our old One Ton Camp with fuel for one hot meal and food for two days. For four days we have been unable to leave the tent—the gale howling about us. We are weak, writing is difficult, but for my own sake I do not regret this journey, which has shown that Englishmen can endure hardships, help one another, and meet death with as great a fortitude as ever in the past. We took risks, we knew we took them; things have come out against us, and therefore we have no cause for complaint, but bow to the will of Providence, determined still to do our best to the last. But if we have been willing to give our lives to this enterprise, which is for the honour of our country, I appeal to our countrymen to see that those who depend on us are properly cared for.
>
> Had we lived, I should have had a tale to tell of the hardihood, endurance, and courage of my companions which would have stirred the heart of every Englishman. These rough notes and our dead

bodies must tell the tale, but surely, surely, a great rich country like ours will see that those who are dependent on us are properly provided for.

Scott read what he had written. It sounded very good. After the catalogue of misfortune plainly told, there came the expression of patriotism, fortitude and acceptance, and as proof of self-sacrificial feeling a turning away at the end from a contemplation of their own predicament to expressions of concern for those left behind. Scott also liked the way that he had alluded to the financial plight of the expedition in his closing words. He gently reminded the country that they had not supported him very generously, covertly pressuring them on behalf of Kathleen and the dependants of his comrades.

It was a job well done and, Scott thought, one which would help him in composing letters to individuals: it had the right feeling, the right balance. Scott felt thrilled, excited and exhilarated in a way that was new to him; he felt the satisfaction of self-expression. Perhaps this is what real writers and artists felt, the beauty of manipulation, first of their materials, then of their audience. Next he would write to the admirals involved with his career and the expedition; he would not have his career blighted even in death. Scott prepared to compose himself again.

• • •

They had a last pathetic meal. Bowers roused himself and with the help of Wilson used the last of the oil to heat water for a final weak and lukewarm cup of tea. Each of them had part of a biscuit left which they nibbled disconsolately.

There was, of course, more conversation about what was to be done. Bowers was for having a sleep, and then turning out whatever the conditions were like and making for the depot, with he and Wilson dragging Scott on the sledge. Wilson said this was impossible. He no longer had the strength to drag Scott. Scott also objected. If they were all lost in their tracks, then maybe the records and letters he'd written, not to mention the geological specimens, would all be lost, buried in the drift. If Wilson and Bowers went, they should leave him behind in the tent.

But there was only one tent. This meant that Wilson and Bowers would have to walk to the depot and back without a shelter for rest and recuperation. It seemed an impossible task. On the other hand, just to lie there seemed neither becoming nor heroic. The decision to start into that comfortless world was, however, easy to delay.

Scott decided that his last writing for now would be a diary entry. Then with the tea, and a little biscuit in his aching belly, he would attempt to sleep.

> *Blizzard bad as ever—Wilson and Bowers unable to start—to-morrow last chance—no fuel and only one or two of food left—must be near the end. Have decided it shall be natural—we shall march for the depot with or without our effects and die in our tracks.*

That had a fine, heroic sound to it. He had covered the blank spaces of so many pages today with the meandering marks of his pencil. Like the tracks on the great continent that stretched away from their tent, he had made his mark on another whiteness. It comforted him. The words of a poem his mother had read to him as a boy came to him like a talisman: *Lives of great men all remind us/We can make our lives sublime,/And departing, leave behind us/Footprints on the sands of time.* He did not pause to consider how quickly his tracks were obliterated from the barrier surface by wind-blown snow, just as footprints are washed out by the inexorable tide.

Saturday 24 March–
Wednesday 28 March, 1912

WE MOVE DEEPER INTO THE darkness and the mysteries. Like the cairns Scott used to mark his route back from the Pole, his diary entries are a way of providing a track through the wilderness, the unknown spaces of another man's life. Now, however, we lose those bearings. For between 24 March and 28 March, Scott wrote nothing in his diary. Most of the letters are not dated, although there is one to his agent in New Zealand dated 24 March. There is another lucid diary entry on 29 March, so it is possible that he wrote letters on all the intervening days.

As for Bowers, he wrote his last words to his mother on 21 or 22 March. His letter oscillates disconcertingly between admitting the certainty of his own death and the idea that he is still strong and will struggle forward to the end. Wilson also wrote letters on or about those dates: one to his friend (and patron of the expedition) Sir Reginald Smith, another to his wife. Both of these speak of attempting to get to the depot and back, and both forecast the impossibility of achieving this task. To Smith he wrote, 'We shall make a forlorn-hope effort to reach the next depot tomorrow, but it means twenty two miles [there and back], and we are none of us fit to face it.' To his wife more bluntly, 'Today may be the last effort. Birdie and I are going to try and reach the depot 11 miles north of us and return to this tent where Captain Scott is, lying with a frozen foot ... I shall simply fall and go to sleep in the snow, and I have your little books with me in my breast pocket ... Don't be unhappy all is for the best ...'

But in the end they were incapable of making this last gesture. Instead they stayed in the tent and waited for death. By 24 March, with no fuel and little to no food left, they had capitulated. Apathy, one of the symptoms of scurvy, exposure and exhaustion had won through.

· · ·

Lost. Lost in the vast white sheet that was the ice-barrier. In the dream Scott was stumbling forward through the drift but had lost all bearings. He could see no cairns. His companions were no longer with him. At the same time fragments of his letters drifted through his consciousness: 'I shall not have suffered any pain ... we will die in our tracks ... It isn't easy to write because of the cold ...' Ghostly figures appeared like mirages before him. Kathleen with the boy, his friend James Barrie, his mother and sisters, Admirals in cocked hats and plumes. To each he tried to speak. He opened his mouth but the blizzard blew his words away, and the figures passed unheeding. As they disappeared he was on his knees in the snow, struggling with pencil and paper to write messages so that he could run and deliver them to the departing figures. He filled sheets and sheets of paper but nothing made sense. It was as incoherent and meaningless as his movement through the landscape. He was crying out to be heard, but surrounded by silence. His life, a white silence.

He woke screaming. His companions slumbered on. Fragments of the dream came back to him and he wondered about loneliness. After all, here he was lying between two men whom he admired as much as any and yet what did he really know of them? Their physical presence comforted him but the growing silences between them seemed to speak of infinite distances that could not be bridged. Bowers seemed such a straightforward character. A man of action and decision. Things were either right or wrong. The Bible was his guide, accepted unequivocally. His view of life was simple, his words always to some purpose. Now, when action was not possible, he seemed to be sinking into a wordless acceptance in which speech was largely redundant. Earlier Bowers had told tales of his childhood, his sisters and his mother, but now as if he had already said goodbye to them, his stories had ceased.

Wilson too was quiet. Whenever he spoke, it was to express his faith in God, and to share this with his companions. Bowers always assented, but Scott found such faith more difficult. He wanted to believe in God, just as he wanted to believe there was some purpose in their lives and deaths. But the tone of Wilson's and Bowers's acceptance, their passive faith, was something he could not share. He wanted still to control his own fate, his own death, his own memorial.

Scott lit the lampwick again and watched that little light spread its fragile suggestion of warmth and comfort.

He nibbled another opium tabloid, then began to write. He would polish off Bowers's mother and then Oriana Wilson. It was like writing dutiful thank-you letters as a child after Christmas, he thought with a grin. Even then the trick was to say nice things about the useless gifts. Now it was even easier. For there was something real to say, and he believed he had gained a more artful command. Language, he was learning, could be manipulated to conceal far more than it revealed, to both make and mask, to create illusions and shield truths.

It was easy, for instance, to construct Bowers and Wilson as paragons of virtue. This is what their loved ones would wish to hear, and in saying what they wished to hear, Scott would in turn be well thought of, thereby decreasing the chance that he would be blamed for the demise of son and husband. Never mind that somewhere, secretly, he was beginning to admit that he shared some of his wife's impatience with Bill's religiosity, and that he had found at various stages along the way Birdie's optimism facile, his responses to the world crude.

Of course he depended upon them both and so had repressed such observations. But on more than one occasion, when Bowers had responded to some debate with his black and white certainties, Scott had remembered the words of a New Zealand farmer at one of the farewell dances in Christchurch. 'You couldn't kill that bloke if you took a pole-axe to him.' Yes, Birdie had a certain coarseness of mind that went with his strength of body. He was the least sensitive of all of them to the cold, and Scott was tempted to think he was the least sensitive of them in other matters as well. As for Bill, sometimes it was true that Scott found comfort in his faith. Yet at other times the blandness of Wilson's faith irritated Scott's temper. Surely if there were a God, He had a temper too, rather than the mild, diplomatic, sweet reasonableness of Bill.

Nevertheless, it wasn't difficult to address Mrs Bowers:

> I write when we are very near the end of our journey, and I am finishing it in company with two gallant, noble gentlemen. One of these is your son . . . I appreciate his wonderful upright nature, his ability and energy. As the troubles have thickened his dauntless spirit ever shone brighter, and he has remained cheerful, hopeful and indomitable to the end.
>
> The ways of Providence are inscrutable, but there must be some reason why such a young, vigorous and promising life is taken.
>
> My whole heart goes out in pity for you . . .

And to Oriana Wilson:

> I should like you to know how splendid he was at the end—everlastingly cheerful and ready to sacrifice himself for others, never a word of blame to me for leading him into this mess. He is not suffering, luckily, at least only minor discomforts.
>
> His eyes have a comfortable blue look of hope and his mind is peaceful with the satisfaction of his faith in regarding himself as part of the great scheme of the Almighty. I can do no more to comfort you than to tell you that he died as he

lived, a brave, true man—the best of comrades and staunchest of friends.
My whole heart goes out to you in pity . . .

• • •

Scott's 'whole heart' was easily given away in more than one direction. The tenor of these letters begins to become familiar. Physical and mental suffering are denied, while much is made of the aristocratic ideal of the gentleman and the chivalric ideal of self-sacrifice. Providence and God are called upon to exonerate Scott from blame and to dole comfort to the bereaved. Scott disarms any possible criticism from Mrs Wilson in his letter to her by declaring that Bill doesn't blame him for the situation. There are no complaints, only celebration.

• • •

Scott imagined with satisfaction the tearful response of Mrs Bowers and Oriana Wilson to his words about their relatives. It made him think again of those mysteries, love and friendship. Kathleen flitted in and out of his thoughts, just as she had moved in and out of his life—an unpredictable, radiant, if bewildering guest. And then there was Barrie. How peculiar it was, Scott considered, that he had become friends with such a writer. How different Barrie was from the men with whom he now shared his fate. His relationship with them was to some degree understandable; they shared a predilection, an ability for action. They shared conventional ideals. But Barrie was like a creature from a different sphere. Even his physical appearance somehow suggested the elfin imagination—that short, slender frame, the pointed features, the huge moustache. He was both child-like and adult at once, whimsical and serious by turns.

They had met after Scott returned from the *Discovery* expedition and was being fêted all over London. It was pleasant to recall those glittering dinners that could be enjoyed in memory without the shyness, the awkwardness, the occasional embarrassment that overtook him at the events themselves. There was one occasion when he had shamed himself by arriving with an overcoat on and, when it was removed by a butler, it was revealed that, lost in a daydream, he had forgotten to put on his dinner jacket. He stood there in his shirt and braces, surrounded by other guests' laughter, full of confusion as the clipped accents of the smart set chivvied and remarked. He had had to retreat by cab, offering his apologies.

It was typical of him to let defeats like this intrude on otherwise pleasant recollections. He forced himself to recall the dinner when he had first met Barrie. It was at Mabel Beardsley's house on one of the many occasions she invited him. Henry James had been there, and Pauline Chase, who had played Peter Pan so beautifully. It was at just such an evening that he had met Kathleen for the first time, but she wasn't there on this occasion. He had only had eyes for Pauline, whose American accent had charmed and flattered him. He allowed himself to wonder what might have been if he hadn't met Kathleen. Pauline, of course, was not as strong a character as his wife, but she may have made a more biddable partner. But such speculation was useless now. The choices made were made forever; the chances lost were gone.

He forced his thoughts back to Barrie. Before leaving England, Barrie had manifested an inexplicable coolness towards him. But then Scott knew the playwright was an intensely sensitive individual who was given to jealousy over his friends. At the time Scott had been too busy with preparations and raising money for the expedition to be able to expend energy on healing the slight breach in their relations. But now something had to be done to give Scott peace of mind. Barrie, he realised, meant a great deal to him.

His mind returned to their first meeting. The whole evening had been wonderful. He'd had an exciting sense of being at the heart of literary London, and he loved to hear the sparkling conversation of the writers and artists gathered round the table. Though they often made him feel staid and conventional, just to listen to them discuss the questions of the day was a liberation.

Barrie and he had walked home together through the London streets, arguing about the life of action versus that of contemplation. The deserted streets had echoed their footsteps and their passionate debate. Scott had argued that the life of action was not as glorious as it was made out to be. Barrie argued that it was the real test of manliness. Scott demurred. It wasn't the action itself that mattered but what was made of the action for others to enjoy, emulate, respect. And it was the writer's work in the end that made action meaningful. Barrie replied, 'Quite so. Action precedes contemplation and in that action a man finds himself in ways that are impossible sitting at home.' They were each other's alter ego. Scott wanted to be a writer like Barrie. Barrie wanted to be an adventurer like Scott.

This thought inspired Scott to begin his letter. Surely, whatever the rift between them, Barrie would forgive him now that he was dying an adventurer's death:

My Dear Barrie,

We are pegging out in a very comfortless spot—Hoping this letter may be found and sent to you I write a word of farewell—It hurt me grievously when you partially withdrew your friendship or seemed so to do—I want to tell you that I never gave you cause—If you thought or heard ill of me it was unjust—Calumny is ever to the fore. My attitude towards you and everyone connected with you was always one of respect and admiration—Under these circumstances I want you to think well of me and my end and more practically I want you to help my widow and my boy—your godson—We are showing that Englishmen can still die with a bold spirit fighting it out to the end. It will be known that we have accomplished our object in reaching the Pole and that we have done everything possible even to sacrificing ourselves in order to save sick companions. I think this makes an example for Englishmen of the future and that the country ought to help those who are left behind to mourn us—I leave my poor girl and your godson, Wilson leaves a widow, and Edgar Evans also a widow in humble circumstances. Do what you can to get their claims recognised. Goodbye. I am not at all afraid of the end, but sad to miss many a humble pleasure which I had planned for the future on our long marches. I may not have proved a great explorer, but we have done the greatest march ever made and come very near to great success. Good-bye my dear friend . . .

Barrie would appreciate these sentiments—the patriotism, the self-sacrifice, the bravery. Scott could not help thinking of *Peter Pan* after whom his son had been named. Pauline had been beautiful as the boy who couldn't grow up—she embodied the kind of purity that Barrie must have had in mind. And how memorable the scene in which she, as Peter, marooned on the rock in Black Lagoon, said in her bell-like voice, 'To die will be an awfully big adventure.' Scott had felt a thrill of recognition. And later in the play, he had also loved the solemn moment when Wendy, in her role as 'mother' to the lost boys, says some last words before they are made to walk the plank by Captain Hook: 'We hope our sons will die like English gentlemen.'

Here's what he had to do: to make it clear in his letters that he and his companions would die in a manner befitting men of England. He would make himself into the hero of an adventure story.

• • •

Scott stirred from a restless slumber and became aware of his body again. It ached, but not now with the desire that had once so tormented him. He no longer had to deal with the restlessness of his longings. Nor was he troubled by the need to urinate or defecate. As fuel ran out, and as the food was finished, the body no longer made those implacable, undignified and dirty demands.

And with the small doses of opium, for short spells he was allowed to float free of the dying cries of his physical self; it was a liberation that filled Scott with strange exhilaration. All his life he had been tortured by his sexuality, with its ambivalent yearnings and demands. It seemed suddenly the source of all his problems. And now as his body died those problems receded, and he was able to express his sense of fulfilment in and through his writing. He took up his letter to Barrie again:

> *Later—We are very near the end, but have not and will not lose our good cheer. We have had four days of storm in our tent and nowhere's food or fuel. We did intend to finish ourselves when things proved like this, but we have decided to die naturally in the track.*
>
> *As a dying man, my dear friend, be good to my wife and child. Give the boy a chance in life if the State won't do it. He ought to have good stuff in him—and give my memory back the friendship which you inspired. I never met a man in my life whom I admired and loved more than you but I never could show you how much your friendship meant to me—for you had much to give and I nothing.*

Scott re-read what he had written. He was light-headed and grinned as he realised its meaning. After all, it was true. The storm had been in the tent as well as outside it, and they had been eating nowhere's food cooked on nowhere's fuel. Nowhere. Never Land. No body.

Scott did not know whether to feel triumph or defeat. He had at least, at last, admitted his love for Barrie. But this was mixed with envy for his genius and fame. Barrie had spun fictions from what seemed a relatively ordinary life. Whereas he, Scott, had only found words by plodding through this white wilderness in order to be somebody. That Barrie could have something to say without the pain that Scott was enduring seemed unfair. But that was genius. And fate might still snatch the words he'd just written away. What if they were not found? The name Robert Falcon Scott would become a silence, a space, a blank page, rubbed out as surely as his footprints across the great continent that howled its presence outside the tent's fragile wall.

'Still writing, sir?' It was Bowers's voice, at which Wilson also stirred. The three of them, propped up by weakening arms, looked into each other's eyes and read the same sad tale of emaciation and decay, their skulls beginning to peep from the skin. They curled back into their bags.

'What will we have to eat when we get to Hut Point?' asks Bowers. It's a game they've played before, and now Wilson recognises the boyish simplicity of Bowers that masks despair.

He answers as naturally as he can, 'Oh, bread and jam, I should think, like we had when we got back from Cape Crozier with the penguin's egg.'

'No,' says Bowers, 'it'll be better than that—more like the Owner's birthday dinner. Penguin steak, and roast spuds and peas with champagne to drink and the hut decorated with flags. And after dinner we'll be tipsy and dance and have cags.'

Silence greets this outburst. They know he is lost in imagination. Scott and Wilson find nothing to say.

Bowers, sensing the atmosphere says, 'It's nearly over, isn't it, Teddy?'

'Yes, it's nearly over. We'll just fall asleep.'

Anxious for reassurance, Bowers murmurs, 'It's nothing, is it?'

'It's something and nothing,' says Wilson. 'You must have faith. We'll all meet again when this is over, and see all those we love perfected in His sight.'

'Let's sing,' said Bowers. Wilson began with his favourite, 'Lead kindly light, amid the encircling gloom.' They knew they would march no further.

• • •

Scott knew he had to finish the letter to his wife. Strange to think that it would be the last of the hundreds that he'd written to her during the separations which had been their courtship and marriage. He had never spoken or written so much so intimately to anyone. Not only distance lay between them but words, words, words. Scott felt dizzy thinking of them all, trying to explain himself, to express what he felt, to capture the essence of himself, and, more pertinently, to capture her.

When he first saw her, he had been struck by her animation, her quick speech and gestures, her magnificent hair. She was handsome rather than beautiful. Statuesque. With an intoxicated gaze he'd seen the flush spread from her face to her neck and the promise of her breasts as she noticed him asking his neighbour at the dinner table who she was. Kathleen had been seated between Barrie and Henry James, and was obviously giving those two lions as good as they gave.

It was some time later when they met again, and this time he had talked to her. She was the embodiment of everything that Scott felt was missing from his life. Twenty-six years old, a sculptor who had worked in Rodin's studio in Paris, she spoke to him of 'vagabonding', by which she meant walking and living in the open air close to nature and away from the shackles of society. Scott had walked her home that first night and they had talked and talked and laughed and jostled. He remembered the bump of her hips against him, and the curious thrilling enticement of her. Unlike most of the girls he had known, Kathleen had few shynesses or inhibitions. She talked to him like a man. Used to the salons of Paris, she had no time for simpering or false modesty. She disliked suffragettes, believing that most women were only fit for the domestic sphere, and either too silly or not informed enough to be trusted with a vote.

Scott had attempted from the beginning to express his attraction to the kind of life she'd led and was leading. He spoke of his first expedition to Antarctica, the way he had felt freed from the money-grubbing aspect of society, and how he loved the purity of that white landscape. It was, he averred, a kind of disciplined vagabonding. But even as he'd spoken he knew that the navy, the service that was his life, demanded conformity from him. Would this wonderful, romantic, artistic woman ever be able to look at his conventionality with anything but disdain? And then there was the problem of his mother, whose financial security could not be compromised, and to whom he felt an enormous emotional obligation.

Could he afford to court and marry Miss Bruce, who had little money to her name, though she was the orphaned daughter of a respectable clergyman's family?

But Kathleen had in her breathless, enthusiastic way encouraged his attentions. From the first Scott both loved and feared this enthusiasm. When it was directed towards him and his plans it was like basking in the sunlight; when her attentions were directed elsewhere, Scott felt the chill. He desperately wanted not to be possessive, not to make himself ridiculous, not to lose his dignity. But when she spoke of her other friends, among whom there always seemed to be an inordinate number of men, he could not still the voices in his head, fearful and warning: *She had, how shall I say, a heart too soon made glad.*

Nevertheless he was compelled, ensnared, entranced. He was tugged along by a power he had no means to resist. Just a few days after their first conversation he was like a schoolboy pacing by her window in Cheyne Walk hoping to catch a glimpse of her, then walking back to his room at the Services Club, penning his ardour to her.

Suffering again, Scott thought. The only other time in his life apart from his Antarctic journeys when he wrote fluently was when he wrote to Kathleen. Some of the letters had made him feel proud as he wrote them. More frequently they were expressions of the difficulties and confusions of their relationship. In the eleven months of their courtship he wrote to her nearly every day. He tried to remember what he'd said, what she'd said, for those precious words somehow were the straw and sand, the bricks from which they had made their uneasy home.

Scott struggled. It all seemed past understanding. They had spent so little time together and so much of it seemed so unsatisfactory. He tried to dredge back the lost words. Obsessive fragments began to repeat themselves over and over again as he lapsed into the crepuscular world between full consciousness and sleep wherein all our ghosts come back with their gentle or violent hauntings. 'Dear sweet face ... soft lips—were they smiling—don't tell me it's all fancy—don't fly away—I've this dread that you can't be real ... a catch of breath, extra revolutions within—do you shorten life, if it is told by heart beats ... don't tell me it's all fancy ... I'm half frightened of you ... You shall go to the Pole ... very gloomy tonight ... I've so little, so very little to offer ... too lonely for words without you ... a dreadful want for you what does it mean ... don't want to feel lost without you ... don't fly away ... we're horribly different ... I'm writing stupidly all the time ... too lonely for words without you ... so little to offer ... very gloomy tonight ... Don't let's get married ... frightened of you ... don't fly away ... You shall go to the Pole ... it's all fancy ... writing stupidly ... writing stupidly ... writing stupidly ...'

• • •

Scott began his last letter to Kathleen, addressed 'To my Widow', before the death of Oates, and it was finished in instalments thereafter. It begins with habitual understatement—'We are in a very tight corner'—and goes on to explain that he is writing during short lunch-breaks 'preparatory to a possible end'. If not quite as cold as the −40° degree temperatures in which it was written, the tone is very cool:

> If anything happens to me I should like you to know how much you have meant to me, what pleasant recollections are with me as I depart.
> I should like you to take what comfort you can from these facts also. I shall not have suffered any pain, but leave the world fresh from harness and full of good health and vigour. This is decided already. When provisions come to an end we simply stop unless we are within easy reach of another depot. Therefore you must not imagine a great tragedy. We are very anxious of course and have been for weeks, but our splendid physical condition and our appetites compensate for all discomforts. The cold is trying and sometimes angering, but here again the hot food that drives it forth is so wonderfully enjoyable that one would scarcely be without it.

The calm assurance of this letter is in startling contrast to the involuted worryings of his love letters, in which Scott is fighting himself, and to some extent Kathleen too. Here he is in control, building a unified image of himself as the strong man, the reluctant hero. These protestations concerning their health, their splendid physical condition and freedom from pain, constitute a massive denial of the realities of their circumstances. What matters to Scott is not to appear weak; the sufferings of the body are denied in favour of moral uplift. Death is reduced to an imaginative possibility, euphemistically referred to as a 'departure', both signifiers as empty as the landscape in which the letter was written.

The letter to Kathleen continues after Oates's demise twenty miles from One Ton Depot. The word 'death' still does not intrude upon the meditation—Oates has simply 'gone'. But now they have 'very little food and fuel'. Scott turns his attention to assuaging his conscience with regard to Kathleen and their son, Peter, whose name is never used:

> I want you to take the whole thing very sensibly as I am sure you will. The boy will be your comfort. I had looked forward to helping you to bring him up, but it is a satisfaction to know that he will be safe with you. I think both he and you ought to be specially looked after by the country, for which after all we have given our lives with something of spirit which makes for example . . . I must write a little letter for the boy if time can be found, to be read when he grows up. The inherited vice from my side of the family is indolence—above all he must guard against that. I had to force myself into being strenuous as you know—had always an inclination to be idle. My father was idle and it brought much trouble.
>
> You know I cherish no sentimental rubbish about re-marriage. When the right man comes to help you in life you ought to be your happy self again—I wasn't a very good husband, but I hope I shall be a good memory. Certainly the end is nothing for you to be ashamed of, and I like to think that the boy will have a good start in his parentage of which he may be proud.

> It isn't easy to write because of the cold −40 below zero and nothing but the shelter of our tents. You must know that quite the worst aspect of this situation is the thought that I shall not see you again. The inevitable must be faced. You urged me to be leader of this party, and I know you felt it would be dangerous. I have taken my place throughout, haven't I ?

The light flickered on the page, making the words dance in front of his eyes. It was more and more of an effort to focus. He had to use his improvised lamp sparingly now, and write as much as he could as quickly as possible, for the spirit was dwindling. But before he could continue he had to read through what he'd written to Kathleen. As he read, he realised he had struck the right note. It was in concert with her beliefs. It would make her admire him—he *would* be a good memory. He was filled with the same poignant sadness that he felt when reading Tennyson or his favourite, Browning. *Let's contend no more love, all be as before love, only sleep.* The bitter-sweet punishment of art.

Just as the storm outside seemed to have abated, the drumming of the tent's canvas muted at last all too late, now the contentions were over for good. Ruefully, he recognised that she would take it sensibly, all too sensibly. And as for her taking another husband, there could be no doubt. He was sure that she had not felt their separations as much as perhaps she might. Her independence of mind and action were in little doubt. Even when she was pregnant she had gone camping with

Paget in Dorset, leaving him to deal with expedition matters in a state of high anxiety. But then Kathleen could not stand his moods, his depression. Try as she might she couldn't hide her impatience. The nagging conviction that she had married him only because of his fame as a polar explorer came back to him with the old sick feeling of vertigo, until he reminded himself that it no longer mattered, and that if this interpretation was correct then she had what she'd always desired.

Certainly she had desired a son. Kathleen had made it plain that it was his privilege to be the father; she was forever emphasising how she had preserved her virginity until a man worthy to play this role was found. Sometimes he wondered if she protested too much. But that was unthinkable. As he had told her in many letters, she was pure of soul when they married. Afterwards he was not so sure. He found it difficult to equate the carnality of the sexual act with purity. Marriage had not liberated him.

It no longer mattered. Another burst of adrenalin from the pain-banishing opium made his spirits soar. He remembered the torments and trials of his honeymoon and felt a strange relief that such embarrassments were over for good. The nerve strain, the fear of inadequacy, the awful self-consciousness with which he had met passion need no longer haunt him. And the thought of being humiliated by Kathleen with other men could now be banished. He could be magnanimous about the idea of her re-marrying; she would be the widow of a hero. No more would he have to feel that he was too constrained for her; no more would he have to endure the icy spaces opening between them. How often had he thought of Browning's lines in connection with Kathleen: *'twas not/Her husband's presence only, called that spot/ Of joy into the Duchess' cheek ... /She had/ A heart— how shall I say—too soon made glad,/Too easily impressed; she liked whate'er/She looked on, and her looks went everywhere.*

Still, he was the father of her child. He had at least served that purpose. Kathleen wanted a baby, so added to the tension of their embraces was the necessity to plan the optimum times of the month. It reminded him of military manoeuvres, but he played his part. Then he would be at sea again. She would write of her doings, her parties, other people, Isadora Duncan. He felt like a footnote to her life. He played his part.

Then the news of her pregnancy came. He was on HMS *Bulwark* when he received word telling him to 'throw up his cap and shout and sing'. Her words were not difficult to remember. 'Me seems,' she had written, 'we are in a fair way to achieving my end.' And despite his initial enthusiasm in which he'd rolled with Everett on the floor of his cabin, feeling the tickle of the man's beard against his face, excited by the feeling of their bodies tumbling together, his eyes were drawn back later to her words and he felt a familiar unease spread through him. It was surely a case for 'our' rather than 'my' end.

But it was true that he had not felt involved in the pregnancy. He was an interested observer, relieved that he no longer had to fulfil conjugal obligations. When the tiny squalling bundle of flesh had been placed in his arms, he'd not known what to feel—a strange impersonal elation mixed with sadness and terror. Elation because somehow here was proof to the world of his normality, his manhood. Sadness and terror because he did not know how to be a father, how to bring himself into relationship with this helpless bundle of human frailty.

Even now, Scott found it difficult to think of his son as an individual human being. He was rather an idea, the boy, for whom provision had to be made. With this in mind, he nerved himself to write again. After telling her their situation, he continued:

> *I think the last chance has gone. We have decided not to kill ourselves but to fight to the last for that depot, but in fighting there is a painless end, so don't worry. I have written letters on odd pages of this book. Will you manage to get them sent. You see I am anxious for you and the boy's future. Make the boy interested in natural history if you can. It is better than games. They encourage it at some schools. I know you will keep him in the open air. Try and make him believe in a God, it is comforting—*
>
> *There is a piece of the Union Jack I put up at the South Pole in my private kit-bag, together with Amundsen's black flag and other trifles. Send a small piece of the Union Jack to the King, a small piece to Queen Alexandra, and keep the rest, a poor trophy for you—What lots and lots I could tell you of this journey. How much better has it been than lounging in too great a comfort at home. What tales you would have for the boy, but oh what a price to pay—Dear you will be good to the old mother . . . Oh but you'll put a bold strong face to the world,*

only don't be too proud to accept help for the boy's sake. He ought to have a fine career and do something in the world . . .

After a few more niceties, instructing Kathleen to remember him to various friends, he laid down his pencil, shut up the little manuscript book in which he was writing and placed it under his head. He snuffed the light and curled low in his bag, exhausted. He had played his part. He had taken his place.

Thursday 29 March, 1912

Since the 21st we have had a continuous gale from W.S.W. and S.W. We had fuel to make two cups of tea apiece and bare food for two days on the 20th. Every day we have been ready to start for our depot 11 miles away, but outside the door of the tent it remains a scene of whirling drift. I do not think we can hope for any better things now. We shall stick it out to the end, but we are getting weaker, of course, and the end cannot be far.

It seems a pity, but I do not think I can write more.

R. Scott

Last entry.
For God's sake look after our people.

So Scott composed his last diary entry, maintaining the heroic tone adopted in his last letters which had skilfully avoided articulating the truth that they were lying in the tent waiting to die. Here at least he has given up the notion that they are going to die in their traces, but they *are* going to 'stick it out to the end', presumably indicating that they would prolong their suffering for as long as possible rather than taking enough opium to kill themselves. There are no complaints. In this way suffering is ennobled, celebrated, made into an ideal. The diary ends with an intimation of just how weak they are (he can write nothing further) and then the famous last phrase with its implication that Scott's last thoughts were not of himself but of those left behind. This projects Scott as a self-sacrificing hero again, and the adoption of the plural pronoun allows the phrase a wonderful generality of application. In the context of patriotism which Scott has made for it in his last letters, the phrase 'our people' can easily resonate to signify the British people in general, as well as the friends and relatives of the deceased.

. . .

Scott woke to the consciousness of cold and confusion. Remnants of a dream still flitted through his mind. He'd been at a party in some fashionable London drawing room. It was crowded. He was stuck in a corner feeling awkward and alone, but there were too many bodies in front of him so he couldn't move. He was in a cold sweat of fear and apprehension. Faces he knew turned from their conversations and in dumb show waved to him, some solemnly, some gaily, before turning back to resume their lively discussions. No one spoke to him, and although he tried to speak he could not. Barrie was there, and Henry James. Shackleton, Nansen and Mawson. All three smiled as they turned away. His mother, her face a crumpled handkerchief of mourning, his sisters solemn and dignified, Kathleen laughing aloud.

The darkness of the tent, the wind keening, the desolation of it all, slowly entered Scott's numbed mind. He had travelled to the heart of the pure continent, to the heart of whiteness, and found there nothing but disappointment, disillusion, the end of all dreams. Amundsen was there first. Scott remembered his own words: 'My God this is an awful place.'

It was in accord, he reflected, with the rest of his life. Kathleen, whom he had thought was the spirit of the open spaces and of freedom, the embodiment of purity of soul, had likewise in the end proved the impurity of all human endeavours and relationships. He could own her as little as the South Pole. All was lost to the cold wastes.

He looked to his companions. They had not spoken for what seemed like days. Their waking and sleeping had not coincided. Now Wilson and Bowers were exceedingly still. With a feeling of panic he realised that they might be dead. With a croaking gasp he half shrieked, half cried, 'Teddy, Birdie!' But there was no reply.

With what was left of his strength, he pulled himself out of his bag and stretched towards Bowers and Wilson, shaking them, repeating their names over and over. But there was no response. He had reached the final loneliness. He cowered in his bag and thought of his mother. The high-sounding phrases of his letter to her resounded mockingly: 'The great God has called me ... I die at peace with the world ... not afraid ... not unhappy ... I wish I could have been a better son ...' It was not his mother he wanted now. It was the mother of his boyhood. The mother to whom he had run for comfort, against whose bosom he had laid his head to rest. It was the mother he had lost forever at thirteen to whom he could never return.

Light-headed, close to delirium, he yearned for the last embrace of death. He took another opium tabloid on his tongue and chewed, ravenous with desire for oblivion. It made him giddy. He felt the adrenalin rush take him and as it did so with his remaining strength he pulled himself half out of his sleeping bag again and tore at his clothes, opening his breast to the cold. In a paroxysm of longing he threw his arm around Wilson and opened his mouth as if to scream his last devotions, but no sound came. Darkness flooded his mind.

interlude

BETWEEN 17 AND 29 MARCH other players were unwittingly engaged in what was to become the drama entitled 'Scott of the Antarctic', which most would read as tragedy. At Hut Point four men, Atkinson, Cherry-Garrard, Petty Officer Keohane and the Russian dog-handler Dimitri, anxiously awaited the return of the Polar party. Others were across the sea-ice at Cape Evans, while Lieutenant Campbell and his group of six men were over two hundred miles to the north-west at Evans Coves hoping to be picked up by the *Terra Nova*. For all these men this was a stressful time, but particularly so for the four men at Hut Point.

On 17 March Cherry-Garrard, Dimitri and their dog-teams returned from a difficult trip to One Ton depot. They were cold and exhausted. The late autumn temperatures were gruelling even for men on full ration. Atkinson had sent them with some extra food supplies and instructions either to wait at the depot or press on further, depending on the conditions. On no account were they to risk the dogs: these were Scott's orders, as he had planned to use them for further sledging the following year. They had reached One Ton on 3 March and stayed there for a week. The weather had been poor, Dimitri was feeling the cold and complained of weakness in his left side. Cherry-Garrard was not confident about his navigational skills. So they waited, hoping that the Polar party might arrive, but they were not unduly worried since in Cherry-Garrard's estimate they could arrive back at Hut Point safe and sound at anytime until 26 March.

But now, as the days begin to pass, anxiety mounts. The men say little to each other but the atmosphere between them is tense. As they go about their routine tasks, all the time it is as if each of them is listening intently for a sound that will herald the return of their comrades. 'Cherry' is particularly jittery. His recent trip on top of the summer sledging as a member of one of Scott's support parties has left him physically debilitated. He feels weak and sick with worry. Should he and Dimitri have pressed on? He willed Scott and the others onward, waiting for the shouts of greeting that would mean that all was well and they could all go back to England in triumph together.

But no shouts were heard. Instead, one night, all four men woke to a sharp rapping sound. 'It's them, it's them!' Keohane had shouted, and they had tumbled from their bunks and rushed to the door of the hut, only to be confronted with the indifferent gale, the night, and the eerie wastes of ice stretching away into the distance. They had trudged back inside shivering and dispirited. A few days later, they were all inside the hut when they heard the dogs begin to howl. This was usually the sign that someone was approaching. Again the shout went up: 'They're here, they're home!' and again they rushed to the door and peered through the drift into the distance. Atkinson thought he could hear footsteps. But there was no answering call, no movement but the swirl of snow. Nothing but the cold, the desolation. They retreated to the hut and stood together huddled by the stove. Nothing was said.

On 27 March Atkinson and Keohane decided to try sledging towards One Ton depot again. They did not get far. The conditions were so severe that they turned back after only two days. By 30 March Atkinson had given up hope. He felt sure the Polar party must have perished in some unforeseen accident. Cherry continued to hope desperately for several more days before conceding to his diary that he thought there was no chance he would ever see his colleagues again. There was nothing to be done now but to trek back across the sea-ice to Cape Evans and prepare themselves for another winter in a hut with five empty bunks to remind them constantly of their losses.

• • •

27 March was Kathleen Scott's birthday. In London she recorded in her diary the success of a birthday party that had been attended by one hundred and ten guests. For over a month now she had been waiting for news of Scott's reaching the Pole. Instead she heard of Amundsen's success, but nothing of her husband. On 11 March, after learning of the Norwegian's triumph, two-year-old Peter asked his mother if Amundsen was a good man.

'Yes, I think he is,' Kathleen said.

Then Peter said, 'Amundsen and Daddy both got to the Pole. Daddy has stopped working now.'

It had made Kathleen shiver slightly. It was the second time that her baby son had said something ominously portentous about his father's progress. The previous September, with no prompting and apropos of nothing, he had said to her: 'Daddy won't come back.'

Under these circumstances, Kathleen considered that it might seem an odd time to have a party. But there was the principle of the thing. She needed to be true to her creed of joy. What was the point of gloom and anxiety? That would not make news, good or bad, come any quicker. Better to lose oneself in the moment. After all, that is how she had got through the separation.

It had been a rich time for her since her husband's departure, she reflected. As she had written to him at the end of 1911, sending her diary to him via the *Terra Nova*, she had had a very happy year. Her work had been a success. She had earned about three hundred pounds from her sculpture and enhanced her reputation through various commissions and exhibitions. Then there had been the trips to Europe, her times with Nansen in Berlin, her wanderings in the Paris of her youth. And she had returned to more socialising and dancing in London. With her young men friends she danced; her older men friends were for conversation. She felt at ease with men. She avoided the company of other women as much as possible, as she felt little affinity with creatures of her own sex.

And always to keep her anchored to earth, binding the other aspects of her life into a pattern of joy, there was

Peter. She felt for her son emotions that she felt for no other being; he was her miracle, flesh of her flesh, bone of her bone. There was simply no other connection like this. She had come a long way since she had first seen a naked man as an art student in Paris and felt physically sick at the sight. Now, having cast off the shackles of such puritanism, she was inclined to think that the profoundest meaning in life stemmed from the corporeal. In both her art and her life she felt at one with the natural world through her apprehension and experience of the physical.

Such thoughts reminded her of longing for the embrace of passion, for a fulfilment beyond dancing. She thought of her husband with some little apprehension. How much longer would she have to wait for him, and how satisfying would their reunion be? For all this time he had existed to her as an idea or an ideal. It was easy in his absence to remember the hero of London, the charmer with the smile, those eyes so blue they were almost violet. It was easy, too, to think of him returning triumphant from such a great endeavour. What confidence it would give him, and who knows what rank and position he might attain? But now there was uncertainty. Amundsen's triumph could not lessen Scott in her eyes, but the public was fickle and difficult to predict. Even more so was her husband's reaction. He would be gracious in public, as he had always been to Shackleton, but how would he react in private? What if he had not got to the Pole at all?

She tried not to think of his depressions. When she thought specifically like this about him, she became troubled. She remembered with an involuntary shiver the times early in their marriage when he had seemed to retreat from the world into some sullen place in his mind where he lost all vitality and appetite for life and became a brooding, heavy presence to those around him. Would a similar depression assail him on returning to England, knowing that the ultimate prize had just eluded him? How would he settle to life as a naval officer again? How would she settle to the role of wife after these years of freedom? Sometimes she felt that she was a good mother and a moderate sculptor, but she was not convinced about how good a wife she was. She prized her liberty too much.

Kathleen told herself to cheer up. She was behaving like the little women she despised. She *would* find something or someone to amuse her and make her days tolerable until news came. And then there would be the excitement of New Zealand and the expedition's return. Nothing could stop such excitement. It was not as if she was going to be plunged back immediately into domestic life. There would be celebrations and perhaps even dancing to be had before that. With this thought she laid her journal aside and slow-waltzed herself to sleep.

Kathleen would not hear of her husband's death for another eleven months.

...

On the evening of 14 June the men in the hut at Cape Evans engaged in an intense discussion. Atkinson had taken command of the party and had relaxed the naval discipline which had separated officers and seamen under

Scott's dispensation. So all hands were invited to a discussion of what should be done when sledging became possible again in the spring. The problem was that they did not know the fate of Campbell's party. There was a slim chance that the *Terra Nova* had taken them off as she steamed along the coast on her return journey to New Zealand. But if the ship had failed to do this because of the conditions, then Campbell and his men would be wintering as best they could, living off what seal and penguin they could kill. The question before them now was whether in November they should sledge to help the living or to find the dead. Since nine men had already gone home with the ship, there were now not enough men and dogs to provide two search parties.

It was an excruciating decision. On the one hand they could travel south and never find their dead comrades, then return to find that Campbell and his men had perished for lack of help. On the other, they might go north to find Campbell all right, and then never know what had happened to Scott and the others or whether they had reached the Pole or not.

Atkinson expressed his opinion that they should go south. Then asked each of the men in turn what they thought. As one after another said south, the decision must have taken on a momentum of its own. Only one man abstained. All the others voted to search for the dead.

In retrospect this seems an extraordinary decision. They were all prepared to gamble with the lives of Campbell and his men. Nothing demonstrates so strongly that the men felt the whole meaning of the expedition and its reputation depended entirely on the attainment of the South Pole. Despite Scott's insistence that he wasn't going to 'race' Amundsen for the Pole, and the consequent emphasis he placed on 'science', everyone knew that the renown of the expedition depended on knowing whether Scott reached the Pole and on knowing how he and his men had died. To return to England without this knowledge was a recipe for ignominy. Under these circumstances the fantastic risk they took becomes much more understandable.

• • •

They found the bodies on 12 November, having spent the previous night at One Ton depot. As they proceeded, some leading ponies, some with the dog teams, those in the rear saw Silas Wright stop, point and then veer off on his own. As all the parties came to a halt, Wright hurried back breathlessly towards them. 'It is the tent,' he gasped. There was a moment of stunned comprehension. The Polar party had died just half a day's march from One Ton depot. For Cherry-Garrard it was a moment of appalling realisation. He thought of his journey there last March. He had been so close to where they had died. Could he have saved them? Exactly when had they died?

The men hurried over to the tent, which was mostly covered in snow. Only the peak was visible. Several of them began the melancholy task of digging it out, while others stood silently in groups, no one wanting to imagine what was inside. Some of them wept openly. No one had anything to say.

Atkinson was the first to venture into the tent. In the gloom he saw three shapes, two of them prone in their sleeping bags. The third body, which Atkinson soon recognised as Scott, was half out of his bag, with his clothing disarranged and with his arms across Wilson. As he peered at the pinched and frozen faces, he read the terrible marks of suffering there. Before he let himself out into the light and air again, he noticed that all their gear was neatly stowed. Their diaries would doubtless tell the story. All was ship-shape. Only Scott's body out of place.

Before anything was moved, Atkinson ordered each of the ten other men to view the remains. He would not have doubts in anybody's mind about what or whom had been found. One after another they entered the macabre sepulchre. Most of them recorded the scene as 'ghastly'. Some felt that the skin of the deceased looked like old alabaster. To others it was discoloured and yellowing. All of them felt that Scott must have been the last to die.

When they had completed their ritual observances, the gear was removed from the tent, including the letters and diaries written by the three men at the end. Then the tent was collapsed over the bodies, and the whole made into a huge cairn with ice blocks. On the top they fashioned a cross out of skis. Atkinson read the burial service while the other men stood, their heads bowed to the cold, in silent awe at the desolate scene before them.

The search party erected their tents and ate some food, though few of them had much of an appetite. Atkinson read Scott's diaries as quickly as he could, starting from the end and reading backwards. When he had a sense of what had happened he called the men together again. Briefly he told them of the deaths of Evans and Oates, and the story of the final days of Bowers, Wilson, and Scott. He then read to them Scott's description of Oates's death and the 'Message to the Public'. All were profoundly moved. If anyone thought of scurvy or mismanagement, nobody said so. They stood together in those barren wastes, the first of many men to be seduced by Scott's prose, inaugurating the heroic myth of his suffering and sacrifice.

On the depressing journey back to the hut, the minds of the search party must have wondered if Campbell and his men were also going to lend their names to the roll of the dead. But their spirits were lifted and anxieties assuaged when they reached their base to find that Campbell and his contingent had survived the winter and made the five-week march from Evans Coves to the hut safely. They had arrived filthy as coal miners from the blubber stoves that had kept them alive, but they were all in good health and had many stories to tell of their wintering.

Meanwhile, several hundred miles to the north-west, another journey had just begun. That Scott had landed Campbell near Cape Adare—in the vicinity that Mawson wished to explore—had not deterred the young Australian, who had persevered and taken his own expedition to a point he named Commonwealth Bay further down the coast. On 10 November Mawson and his two companions, Belgrave Ninnis and Xavier Mertz, left the winter quarters of the Australasian Antarctic Expedition to sledge as far to the East as they could go in the time available to them. On 12 November they were snowed in by blizzard, but their hopes and spirits were high—this was the adventure they had variously dreamed of—a trek into the cold unknown for science, fame and fortune.

PART TWO

australian engagement: Mawson's Story

Mawson, Douglas 1882–1958

Born Bradford, Yorkshire. Brought to Australia as a child. Educated at Fort Street School and Sydney University. In 1903 he took part as a geologist in an exploration of the New Hebrides under the auspices of the British Commissioner. He became Lecturer in Mineralogy and Petrology at Adelaide University in 1905. Joined Shackleton's 1907–09 Antarctic expedition as scientific assistant to Edgeworth David. In March 1908 Mawson was one of the first party to ascend Mount Erebus, and on 16 January 1909 Mawson, with David and Mackay, was the first to reach the vicinity of the South Magnetic Pole. Mawson arranged and led the Australasian Antarctic Expedition of 1911–1914, which explored and named King George V Land and parts of neighbouring Adélie Land. It was on this expedition that Mawson survived a lonely trek back to base camp following the death of his two companions. He was knighted in 1914. From 1921–52 he was Professor of Geology at Adelaide University. Between 1929–31 Mawson led two further Antarctic trips, carrying out aerial surveys and marine research. He died in October 1958.

This is an account of the last days of his solo march back to Commonwealth Bay in 1913.

17 January, 1913

MAWSON TRUDGED FORWARD. The lowering sky and swirling snow turned the whole of his vista dirty grey. He had been alone for ten days. Physical exertion no longer erased the mind's tormenting return to the fact of his fierce hunger, the torture of his isolation. With every step forward, eyes creased against the drift, he tried to concentrate on maintaining his course. But just as the terrain was a maze of shattered ice and crevasses covered with soft snow which afforded no easy or straight way forward, so his mind wandered from its fixed purpose to become lost in the thought of great plates of piping hot food, juicy steak and vegetables; he dreamed of sitting down to eat in the company of fellow human creatures in the comfort of a room.

Then his body would sharply remind him of immediate, physical realities. Bent double as he hauled his world up a slope heavily covered with soft snow, muscular cramps attacked his stomach and abdomen, tilting him to one side. He felt the peculiar sensation in his boots as the detached soles of his feet slid slightly against their bindings. Despite the lanolin with which he daily anointed the raw parts of his rotting body, with every step his raw scrotum rubbed agonisingly against the wool of his undergarments. In order to concentrate on his course, Mawson tried to play games, telling himself over and over again that all he had to do was to put one foot in front of another and then all would be well. He repeated like a mantra that each step forward was the only way home.

At least it wasn't too cold. *Count your blessings, name them one by one*; they had sung the hymn on Sunday evenings through the long Antarctic winter, cheered by their fire and the hiss of the acetylene lamps while the voracious wind shrieked outside. Now Mawson stopped, took off his mittens and his outer garment and strapped them to the sledge. He pulled again and after a few steps felt terror as his feet dropped through the surface until he was arrested thigh deep by his arms. With considerable effort he pulled himself out and, realising that this must be a snow bridge over a crevasse, peered about him to discern its trend. But the gloom made it difficult to see.

Mawson decided to turn north in an attempt to cross square-on, when suddenly he plummeted downwards with the fearful rush of nightmare. As the rope and harness attaching him to the sledge unravelled, so did his hope. But then he was checked by a mighty jerk which felt as if it might remove his weakened arms. The rope pulled up and he was suspended, slowly revolving, fourteen feet into a giant grave of ice. He felt the sledge tugged by his weight towards the lid of the crevasse. So this is the end, he thought.

But the sledge must have caught on a ridge of ice; it did not follow Mawson into the crevasse and hurtle his battered frame into oblivion. He swung there, peering upwards, watching the rope cut dangerously into the soft snow where he had broken through the lid of the crevasse. On either side of him were blue ice walls which he could just touch if he swung himself slightly; they were about six feet apart. To look down was to see infinity, a black space with seemingly no end.

Shocked and cold as he was, Mawson knew that he must act to survive or die. He considered the impossibility of the task ahead. He was weak and emaciated. As he had fallen, his clothes had filled with snow. He began to feel terribly cold. For days now he had thought that the end was imminent. Was this the form it was to take? He could cut himself free and tumble into the enfolding darkness where there was no more cold, no more suffering. But on the sledge there was still food left; the craving in his stomach commanded a last effort of will. He could not bear the thought of missing the pleasure of that food. Mawson reached above him and felt the rough, hairy texture of the rope against his bare and frozen hands.

There was a knot. He gained some leverage and heaved with all his strength. As he felt his body respond and move upwards, it was as if other hands were lifting him. The Providence that had stopped the sledge was inspiring him. Or so he told himself.

Mawson stretched again. His hands found another knot. He cried out involuntarily with the pain of grasping it; like the rest of his body, his hands were in poor condition, with little skin left on the palms and the fingertips blackening with frostbite. The intense burning sensation of the rope reminded him incongruously of burning his hands and legs on a rope swing when he and his brother were boys in Rooty Hill. But then he had merely fallen into a creek bed. Here such landings seemed soft. Mawson, breathing deeply and shuddering with the effort, hauled again, and again with a groan he felt a response as he moved another body length towards the surface.

After gaining his breath he repeated the process once more, and with a feeling of relief and elation began to thank God, as with his arms and shoulders he attempted to pull himself onto firm ice. But as he pulled, the snow ledge gave way again, and once more he felt the power of gravity pull his stomach up into his chest as he hurtled back down into the chasm.

Like a toy he revolved slowly, almost elegantly, at the end or beginning of fourteen feet of torture. Sheltered from the wind, down here there was an unearthly silence, and the shining blue of the ice walls mocked and beckoned with their indifferent beauty. He thought of cutting himself free from the harness. How good it would be to sleep forever; to know no more of loneliness, pain, the intolerable toil of the march. The temptation was almost overwhelming. Even if he managed to pull himself out of the crevasse, his chances of surviving were slim at best. But before he reached for the knife in his belt, which could sever him from travail forever, another thought intervened.

Mawson remembered Ninnis's fatal fall into a crevasse, and how in looking for their companion he and Mertz had stared down into a seemingly bottomless shaft like the one he swung in now. Although there was no sign of their companion, they could dimly discern, caught on a ledge about a hundred and fifty feet below, one of the dogs crying piteously, its back broken. Thinking of this forced Mawson to consider the possibility that death was not guaranteed if he cut the rope. To drop onto such a ledge and lie with a broken limb, to die such a lingering, terrible death, was worse than the situation he was in. And there was still the food to think of, and the Providence that had saved him thus far.

A final effort was called for. His nerves strained to breaking point, he reached upward again for the knots that would tie him to life. As he hauled himself upwards, his whole body quivering and crying with the effort, he felt a presence with him helping him to endure. It was as if a power not his own infused his limbs with energy. This did not mean his ascent was easy. Several times, having gained the knot above him, his strength gave way and he slid back. But after a massive struggle, he gained the lid of the crevasse again, and this time pulled himself out feet first, using all of his six foot, three inch frame to manoeuvre his legs over the soft snow and onto firm ground.

He lay on his back gasping great lungfuls of the abrasive air, watching the low, dark clouds inexorably moving on above him. Then Mawson shivered and remembered his bleeding hands and his fingers with their blackening tips. He lurched to the sledge, forced his outer windproof shell over his head and put on his mittens. His legs were trembling violently and he felt incapable of further effort. Sinking back down, he lay mesmerised by the passing clouds, allowing his mind to drift with them, wishing he could float home to Australia.

He passed into a numbed prostration from which he was only roused by his own shivering. He did not know how long he had been lying there, but now became aware again of the need for action to save himself from the depredations of the cold and wet.

Thankfully the wind was not blowing hard; Mawson's struggle with his makeshift tent was not as protracted as it might have been. Still it took him one and a half hours of exhausting labour before the miraculous moment when, sitting inside his shelter, the primus was brought to life, and a small glow of light and heat radiated through the gloom. Then there was food. He allowed himself a whole cupful, mixing pemmican with the jelly he'd made by boiling the bones of the last dogs. This was warmed in the cooker and then he added a quarter of a hard-tack biscuit, having hammered it first with his knife handle.

Mawson forced himself to eat slowly. Ravenous as he was, and wonderful as the food tasted, he knew from bitter experience the terrible feeling at the end of a meal when all he wanted was more. So he savoured each mouthful, chewing and chewing for as long as possible.

When he'd done, he crawled into his bag and, warmer

now, took out his stub of pencil to scrawl the journal entry for this most arduous and momentous day of his life.

• • •

On exactly the same day, 17 January, one year earlier, Scott wrote in his journal at the South Pole, 'Great God this is an awful place.' He had been travelling only ten days longer than Mawson, but had travelled over two hundred miles further. But distance should not be the only arbiter in comparing their journeys. For the nature of the terrain, and the circumstances in which it was crossed, must also be considered. There were ample reasons why Mawson did not cover as much ground as his more celebrated colleague.

When Mawson reached his violent hiatus in the crevasse on 17 January 1913, the extremity of his situation is difficult to imagine. He and two companions, Dr Xavier Mertz and Lt Belgrave Ninnis, had set off from the winter quarters of the Australasian Antarctic Expedition at Commonwealth Bay in Adélie Land on 10 November 1912. Their team constituted one of six sledging parties which left the safety and comfort of the hut to explore the environs of Adélie Land to the south, west, and east of winter quarters. Mawson's journey was designed to be the most ambitious. With the help of dog-teams he and his companions aimed to travel further and faster than their man-hauling colleagues in the other parties. Mawson's project was to travel as far east as he could in the time permitting, plotting the coast and exploring the land adjacent to the South Magnetic Pole.

Mawson, Mertz, and Ninnis set off with three sledges and seventeen dogs. Their total load weighed 1723 pounds. The terrain and climatic conditions they encountered were extreme. Their way took them from the hut up a steep ice road for five and a half miles where they reached a plateau, and the previously dug-out ice-cave known to the expeditioners as Aladdin's Cave. From there they proceeded east and discovered that because they were near the coast, the pressure of the inland ice had created a country of massive irregularity. Repeatedly they had to ascend to around three thousand feet and then descend to sea level as they negotiated the mountainous terrain. The ice fields were riven with hidden crevasses; sastrugi hindered their progress. High winds and driving snow made the travelling arduous and navigation perilous.

They crossed two glaciers, naming them after Mertz and Ninnis respectively. In the beds of these massive, frozen rivers, dogs and sledges had to be manoeuvred over crushed, broken and pinnacled ice, while crevasses had to be constantly identified and negotiated. Some were a hundred feet wide, others opened like great cauldrons. Elsewhere the ice was sculpted into towering monoliths, inspiring awe in the three men as they wrestled with the dogs through the tortured and tortuous labyrinth. Then suddenly a snow bridge would collapse and dogs, sledge, and sometimes one of the men would dangle on the brink of infinity until hauled out by their comrades.

Their limbs were lead, their nerves strained to breaking pitch. But they battled on, with Mawson chafing at the slow progress and the fantastic difficulties that they daily encountered. On 14 December they had travelled approximately three hundred miles, and Mawson knew that in a few days they must turn back in order to reach the hut by 15 January. Their plan was to depot some food, make a dash as far to the east as they could go, then turn back, pick up the cached supplies and travel back further inland in an attempt to find better ground for their return march.

But before any of this was achieved, disaster struck. They were travelling over a field of frozen snow known as névé, not expecting any difficulties, when Ninnis, his dog-team and sledge were crashed into oblivion down a crevasse. He had been in the rear, following Mawson, and so was driving the sledge with the most essential supplies on it, as well as the best six of the surviving twelve dogs.

Mawson and Mertz were left three hundred miles away from winter quarters, with ten days' sledging rations, no tent, and six dogs in very poor condition. They improvised a shelter and cooking utensils and set off on the return journey, knowing that they were attempting to cheat death. As the dogs weakened and were unable to carry on, they were shot and used for food. The last dog was killed on 27 December.

Mawson and Mertz man-hauled forward, both men beginning to show signs of severe physical deterioration due to malnutrition. Mertz died on 8 January. Three days later Mawson wrote:

> *My whole body is apparently rotting from want of proper nourishment—frost-bitten fingers festering, mucous membrane of nose gone, saliva glands of mouth refusing duty, skin coming off whole body.*

When he sat in his tent on 17 January to write down his terrifying experiences of that day, he had travelled in total approximately five hundred and twenty miles. He had been alone in the icy wilderness for nine days. He still had approximately seventy-five miles to go before he reached Aladdin's Cave and safety. Given this catastrophic situation, the diary entry is a model of restraint:

> *Overslept after looking out several times during night to find overcast and light snow falling. Got away about 8am. Camped by noon after doing barely 2m on oouroo 30°W of N. Considering all things the sledge was running fairly well, and I intended doing more when the march was precipitately brought to an end.*
>
> *There had been no sun during the morning . . . Light extremely bad and only by great strain on eyes could I keep a course. I escaped several large open crevasses by Providence, not seeing them till past them . . . 'I blundered blindly on' . . . A few moments later I was dangling on end of rope in crevasse, sledge creeping to mouth. I had time to say to myself, 'So this is the end', expecting every moment the sledge to crash on my head and both of us to go to the bottom unseen below. Then I thought of the food left uneaten on the sledge—and, as the sledge stopped without coming down, I thought of Providence again giving me a chance. The chance looked very small as the rope had sawed into the overhanging lid, my finger ends all damaged, myself weak and as it had been warm day I had taken off most clothing and left rest very open—all now filled with snow and*

fingers fast chilling. With the feeling that Providence was helping me I made a great struggle, half getting out, then slipping back again several times, but at last just did it. Then I felt grateful to Providence. I was wet and cold and overcome, so decided to put up tent.

It is impossible to say what is ahead, for the light gives no chance, and I sincerely hope that something will happen to change the state of the weather—else how am I to keep up my average. I trust in Providence, however, who has so many times already helped me.

Mawson is not striving for effect here, but there is an unintentional moment of pathos when he uses the pronoun 'us' to describe himself and his sledge, as if he cannot bear to think of himself as being alone. In the final paragraph there is extraordinary stoicism, even humility, in the phrases 'I sincerely hope' and 'I trust'. Though he was a highly practical man, his mind trained in the sciences, in this predicament Mawson had invented a God to walk with him.

• • •

Mawson drifted in and out of a light and troubled sleep in which vivid dreams painted themselves across his mind. He was walking along North Terrace in Adelaide and as he turned into the university entrance found there, instead of his workplace, a cake shop. Staring through the window he saw a massive cake with white icing. It was not unlike a wedding cake. He went in, but found he had no money with which to buy it. The pretty girl behind the counter firmly refused him credit. Then Paquita was in the tent with him and he was cooking a dog's skull for breakfast and wondering how to divide it in four so that Mertz and Ninnis could have some too. But just as the fragrance of the cooking meat began to entice him, Paquita took the pot and threw out the skull, saying she would not eat such a hideous thing. He scrambled out into the snow to find it, but instead found himself peering over the rim of a crevasse, shouting, 'Ninnis', 'Cherub', 'Ninnis'. He felt the rising panic and desperation as he stared into black silence; Ninnis, food, Paquita, all lost.

As he jolted awake again, he recognised the amalgam of fear and desire from which his nightmares were woven. To be wakeful might be preferable, but his mind would not be still. As if in the grips of a fever he obsessively returned to the horror of his trial in the crevasse. If it happened again, would he have the strength to save himself? There was also the question of lost time and distance. Today he had only travelled two miles—it was not enough. To have any chance of survival at all he needed to do at least five miles each day. If the weather continued badly, he had no chance. If the weather proved favourable, he had very little. What could he do to improve the situation?

The gnawing in his stomach turned his mind back to food. A fragment of Omar Khayyam came back to him:

Unborn To-morrow and dead Yesterday,
Why fret about them if To-day be sweet?

The lines offered a sweet temptation. He could eat as much as he wanted for two or three days, and then make an end of it. He had enough primus alcohol to ease his passage into the next world. Here was a plan of comfort. And it was one that meant he would not have to return to face the tribulations attendant upon the death of his two companions; he would be joined with them in the equality of sacrifice. Death was only a sleep. Had he not always believed as much?

He shivered. The scientist was not the whole man. Death had a peculiar metaphysical frisson for him. There was something terrifying about infinity; there was something abominable in the thought of not-being. And the Persian philosopher was surely wrong. To live in a continual indulgence of the moment meant a denial of consequence; ethics demanded thought for the future in order that action should not be destructive to others as well as wilfully self-destructive. And what if today were not sweet? To eat his fill after all these days of abstinence might make him ill. Then there was Paquita to consider. Mawson tried to conjure his fiancée's face, her warm eyes and lissom body. His memory refused. Yet she existed to him as a feeling of warmth and tenderness. She was, he was sure, both straight and true. Would it not be a betrayal of his promises to give in and leave her so grief stricken so young? Mawson managed a grin at this little piece of vanity. Yet he was sure that she *would* grieve.

The thought of her mourning goaded desire. Although the huge degree of physical and mental stress he was under meant that sexual feelings were as frozen as the landscape through which he daily struggled, nevertheless the promise of consummation meant much to him. He had waited so long that whenever he thought about the physical expression of his passion he was filled with an ardent curiosity as well as a sense of wonder.

All this turned his mind to other lines of verse, these by his favourite poet, Robert Service. The poem was called 'The Lone Trail':

Ye who know the Lone Trail fain would follow it,
Though it lead to glory or the darkness of the pit.
Ye who take the Lone Trail, bid your love goodbye;
The Lone Trail, the Lone Trail follow till you die.

Today he had glimpsed the darkness of the pit, and the reality of nothingness had stared up from the depths of the crevasse. Now it was up to him to take the lone trail to glory. He would prove to be one of the 'silent men who do things' whom Service so admired.

If he was to live he must turn his mind to the practical task. Apart from rationing his food as carefully as possible and praying for the weather to be fair, what else could be done to maximise his chances of survival? Mawson turned this question over and over in his mind. There was a misery in dwelling on the realities of his situation, but he forced himself to the puzzle, knowing that dreams and fine phrases would only take one so far. Action was required. He made a pact with himself to care for his body as much as he could. This would mean making the tedious effort to undress and treat the raw parts with lanolin at every opportunity. He would also endeavour to expose the festering parts of himself to the healing effects of the sun whenever occasion arose. All this meant expending energy, but it was crucial to keep what little physical health he had for as long as possible.

But there was also the nightmare possibility of the crevasses to contend with. Mawson allowed himself to relive the ordeal of extricating himself from the gaping chasm. And suddenly into his mind came the image of a rope ladder. If he had such a thing then he could simply step up out of danger. At last here was something productive for his mind and hands to work on. He had some alpine rope. Mawson roused himself from his sleeping bag and, lighting the stove for a little heat and light, began to fashion himself a rope ladder. If he tied one end to the bow of the sledge and carried the other end over his left shoulder loosely attached to his sledging harness, then if he fell into a crevasse the ladder would unravel and he would simply be able to climb out, like a man climbing the rigging of a ship.

There was satisfaction in the work, and an absorption of his mind in the task at hand. But when the makeshift ladder was finished, and Mawson had retired once more to his bag, dark thoughts came crowding. It was as if two voices assailed him, one soft and alluring, whispering 'Eat and die easily'; the other harsh and grating, rasped 'Starve and live hard'. It was the thought that his struggle might nevertheless end in defeat that was hardest to be borne.

18 January, 1913

FRAYED THREADS OF DREAM still ran through Mawson's mind as he struggled towards full consciousness. Malnutrition, he observed, not only increased insomnia but also seemed to increase the incidence and intensity of dreams. It was 7a.m. From the gloom within his makeshift tent it seemed clear that the light had not improved. The misery of this realisation, and the memory of his cold and broken night spent shivering and tormented with anxiety, decided Mawson on the necessity for at least having breakfast, however little it might be.

With the primus lit, and some snow melting for a cup of hot water flavoured with a half-teaspoon of cocoa, Mawson made himself look forward. There were plenty of hours left today for the sky to clear. At the earliest opportunity he would get away, and his new rope ladder would be his insurance and talisman.

When the cocoa was ready, it tasted so thin and insipid that Mawson could not resist stirring in a little pemmican and broken biscuit to make it more substantial. As he did so it was difficult not to think of his erstwhile companions, now lying in their icy graves. Mertz and Ninnis had been the best of friends; their chatter and mutual affection had cheered him along the way, although he as leader of the party and the expedition had maintained a distance from both of them. But sledging together inevitably encouraged certain kinds of intimacy. And when Ninnis was lost, Mawson felt the need to grow closer to Mertz.

Now, as he used the spoons that Mertz had carved from fragments of the broken sledge in the aftermath of Ninnis's death, Mawson was also reminded of the wonderful breakfast they had shared from Ginger's skull. Ginger had been the last dog to die and, after they'd shot him, they'd sat in the tent for hours cooking the meat and bones. The following morning they feasted, boiling the skull whole in some melted snow over the primus. Fragrance filled the tent. Mouths watered. Mawson and Mertz grinned at each other in anticipation and avidly discussed the fairest way to divide the meat. They decided to score a line down the middle of the skull, and then play 'shut-eye' to determine who should have which half.

When the meat was cooked there was a further tantalising delay while it cooled enough to be handled. Once they managed to lift it from the pot, however, it didn't take long for the temperature to drop. Mawson shut his eyes and turned his back to the skull, lying on its pannikin with its blank white eyeballs showing starkly in the gloom. Mertz pointed to the left-hand side and said 'Whose?' Mawson replied 'Yours'. And so the meal began. Despite the fact that Mertz was not a particularly keen meat eater under normal circumstances, on this occasion he went to it with a will. Ravenously both men tore the grey meat from the bone and then passed the skull, juice dribbling down their mouths and fingers. Ginger had been their favourite dog. They had felt sad and that much more alone when they had had to shoot him. Now, strangely, their love for the animal was no less. They passed the skull and supped from it with reverence. It became a chalice, signifying life and hope.

Mawson was dimly aware of such feelings while simultaneously registering the grotesque contrast between this atavistic ritual and the polite dinner parties of London and Adelaide. He remembered dining with the Scotts in Buckingham Palace Road. The elegance of their dining room came back to him, candlelight playing on the dark and polished wood of the furniture, the silver and crystal glinting in warm light. But no food at that table had ever tasted as good as this. He wondered if Scott had been reduced to such extremities on his way to the Pole.

They ate everything. Mawson was surprised how good the eyes tasted. When each had devoured their respective side of the face and jowls, they laid the skull back on the pannikin, and took it in turns to scoop out the brains and the thyroid with the wooden spoon. After the meal they felt strong and powerful with new energy; hope became easy.

Remembering all this now made Mawson feel empty and alone. The magnitude of the journey ahead threatened to overwhelm him. He had to force his will to dominate fear with rational argument. He told himself that strength of mind would win, and that if he could do five miles a day there was no reason why he should not live to take his place in the candlelight again.

Despite his loneliness, reason told him that the death

of Mertz was a liberation as well as a loss. It was only a day after their splendid breakfast that Mertz had begun to feel ill. Then came a harrowing week in which Mawson had to witness and nurse the physical and mental decay of his companion. They only travelled forward on two days that week. Added to the strain of dealing with delirium and madness, of continually cleaning shit from Mertz's body, there was the stress of knowing that to be eating rations, however little, without making ground, was also the road to his own death.

Mawson shuddered as he recalled those days, and the horrible relief when Mertz had died. With only one mouth to feed, it was still possible to think of survival. And it was to survival that he must now turn his concentration. As soon as there was any light, he must be ready to move. He began to pack up his gear and prepare himself to take up the struggle once more.

• • •

Mawson's account of the deterioration and death of Mertz is stark and spare. On 30 December, he first mentions 'Xavier off colour.' He goes on: 'All his things very wet, chiefly on account of no burberry pants. The continuous drift does not give one a chance to dry things, and our gear is deplorable.' Mertz's burberry pants were lost in the glacier with Ninnis. Mertz was wearing thick woollen long-johns instead, which clearly did not protect him from the wet and cold.

The following day Mawson records that they are 'Keeping off dog meat for a day or two as both upset by it.' On 1 and 2 January 1913 they are prevented from travelling by poor weather. On 3 January they gained five miles, and Mawson alludes again to Mertz's condition: 'Wind frost-bit Mertz's fingers and he is generally in a very bad condition. Skin coming off legs etc.—so had to camp though going good.' There was no progress on 4 and 5 January and the situation becomes desperate: 'I tried to get Xavier to start but he practically refused, saying it was suicide and that it much best for him to have the day in bag and dry it and get better, then do more on sun-shining day. I strongly advocated doing 2 to 5 miles only for exercise even if we could not see properly.' But Mawson could not persuade Mertz, who by now sounds as if he is on the brink of giving up entirely and taking the option of a quiet death in the sleeping bag. 'All will depend on Providence now,' wrote Mawson, 'it is an even race to the hut.'

Mawson's iron will and self-control would not allow Mertz to give in. On 6 January they set off once more despite 'Mertz not being able to help at all'. As the situation worsened, Mawson's diary entries became increasingly lengthy. His account of this terrible day continues:

Surface slippery, so occasional falls. Quite dizzy from long stay in bags, I felt weak from want of food. But to my surprise Xavier soon caved in—he went 2 miles only in long halts and refused to go further. I did my best with him—offered to pull him on the sledge, then to set sail and sail him but he refused both after trial. We camped. I think he has a fever, he does not assimilate his food. Things are in a most serious state for both of us—if he cannot go on 8 or 10 m a day, in a day or two we are doomed. I could pull through myself with the provisions at hand but I cannot leave him. His heart seems to have gone. It is very hard for me—to be within 100m of the Hut and in such a position is awful . . .

A long and wearisome night. If only I could get on. But I must stop with Xavier, and he does not appear to be improving—both our chances are going now.

Although it was harrowing, the following day relieved Mawson of his dilemma. Mertz died, but not before a terrible struggle:

Up 8am, it having been arranged that we should go on at all cost—sledge sailing, I leading and Xavier in his bag on the sledge. Just as I got out at 8am I found Xavier in a terrible state having fouled his pants. He must be very weak now for I do up and undo most of his things now and put him into and take him out of the bag. I have a long job cleaning him up, then put him into the bag to warm up. I have to turn in again also to kill time & keep warm—for I feel the cold very much now. At 10am I get up to dress Xavier & prepare breakfast but I find him in a kind of a fit & wrap him up in the bag & leave him— obviously we can't go on to-day, and it is a good day though bad light, the sun just gleaming through the clouds. This is terrible. I don't mind for myself, but it is for Paquita and for all others connected with the expedition that I feel so deeply and sinfully. I pray to God to help us.

I cook some thick cocoa for Xavier & give him some beef tea—he is better after noon but very low. I have to lift him up to drink. During the afternoon he has several fits & is delirious, fills his trousers again and I clean out for him. He is very weak, becomes more and more delirious, rarely being able to speak coherently. He will eat or drink nothing. At 8pm he raves and breaks a tent pole. Continues to rave & call 'Oh Yen, Oh Yen' for hours. I hold him down, then he becomes more peaceful & I put him quietly in the bag. He dies peacefully at about 2am on morning of 8th.

Death due to exposure finally bringing on a fever, result of weather exposure & want of food.

He had lost all skin of legs & private parts. I am in same condition & sores on finger won't heal.

Mawson uses the present tense through most of this, using a narrative method beloved of adventure writers who attempt to create the illusion of simultaneity between the writing and the events they describe in order to highlight and dramatise the action. It seems likely, however, that Mawson wrote this in stages during the day and into the night and that his adoption of this manner of writing is an unselfconscious imitation of writers like Robert Service whom we know Mawson admired.

The determination to focus on event and action here, rather than to describe any detailed emotional reaction, is a defence mechanism on Mawson's part. But his readers are invited to enter the extremity with our own imaginings. How might we react to the horror of the circumstances? To nurse someone in such straits in the relative comfort of home or hospital is a considerable trial. To so attend a dying friend or colleague in a space barely large enough for two sleeping bags, accompanied by the privations of cold and malnutrition, threatens to overwhelm us.

Despite Mawson's lack of self-reflection, he does allow himself to admit with powerful understatement: 'This is terrible.' He seems to refer to the possibility of his own demise, and we discern the voice of a heavily judgmental consciousness in the assertion that he does not care about himself but feels 'sinfully' towards others.

In this, and other ways, Mawson's puritanical ethics were differently inflected from Scott's. In Scott's last writings there was a deliberate deflection of responsibility, a tendency to blame his and his companions' plight on malignant fate rather than poor judgment. We noticed Scott's shrinking from the body; the way his diary grandiloquently reaches towards the metaphysical while denying, omitting, avoiding, the physical realities of his situation. We saw him courting an audience. Mawson is much less inhibited, much more practical, and much less concerned to convince an audience of his own heroism. Considerable attention is paid in plain language to the physical circumstances of Mertz's death and his own predicament. In this way Mawson enabled himself to survive the horror of his situation.

• • •

Interviewing him in London, Mawson found Mertz to be a man of 'excellent physical and moral qualities and of high education'. He was also of independent financial means. Mawson enrolled Mertz in the expedition immediately and despatched him to Monaco for several months to learn how to use the special oceanographic equipment being lent to the expedition by Prince Albert I. Subsequently, Mertz returned to London and with Ninnis was put in charge of the Greenland sledge dogs on the *Aurora*'s voyage to Hobart. They were the only two of the expeditioners who sailed on this initial stage of the journey to Antarctica. As the rest of the men were Australians or New Zealanders, they joined the ship at Hobart.

Mertz was born in Basel in Switzerland where his father, Ignace Emile Mertz, was a wealthy manufacturer. His mother Josephine was a stiff-backed Swiss-German Protestant. The family were members of the Burgergemeinde, a community of old Basel families who maintained conservative patterns of behaviour and attitude through many generations. One such tradition was that their diet was largely vegetarian.

In his letter of application for the expedition Mertz speaks of his academic and sporting prowess and of his managerial experiences working for his father's manufacturing company. He was a Doctor of Laws and a keen mountaineer. In 1908 he had won the Swiss cross-country ski championship. What he did not confide to Mawson were the recent troubles in his family. By 1910 his parents' marriage had deteriorated to the point where Madame Mertz, supported by her children, brought a court action against her husband alleging that he was insane and not in a fit state to run the business or the family finances. This action was unsuccessful, but Mertz's parents were never re-united. As if to escape the family difficulties, Mertz volunteered for Mawson's expedition in the year that the court case concluded. In his letter he merely speaks of his father having now recovered from a recent illness.

On the voyage from London to Hobart, Mertz and Ninnis forged the beginnings of their close friendship, and both eventually gained praise from the ship's Master, J.K. 'Gloomy' Davis. In keeping with his nickname, Davis initially confided to his diary that Ninnis and Mertz were lazy and useless, but further experience allowed him to come to a more positive view. For Mertz was not only physically strong and courageous but also charming and intelligent. Aboard ship he quickly learned to 'hand, reef and steer' with the best of them and enjoyed going aloft for exercise. In his diary he records that during the voyage he is fully occupied as sailor, dog-handler, ship's doctor and private scholar and claims that he is studying geology so that in the polar regions he might identify where the gold and jewels are situated.

During the winter in Antarctica, Mertz gained the regard and respect of his comrades for his boundless energy, his willingness to help with any task at hand, his sensitivity to the moods of others, and his tendency to express his emotions more openly and passionately than his Australian colleagues. He could speak French, German, Italian and English and brought to the expedition a cheerful disposition and an intense belief in the value of the adventure they had all undertaken.

He had a particularly close relationship to Ninnis and

an attitude towards the younger man that one of their fellow expeditioners described as 'maternal'. Ninnis reciprocated Mertz's affection but, in his English way, was less demonstrative. The two were always considered a pair by the others: it was always 'Mertz and Ninnis' or 'Ninnis and Mertz'.

Despite the homoerotic overtones of their friendship, Mertz also had a reputation for his heterosexual enthusiasms. John Hunter, one of the biologists on the expedition, had two photographs by his bed of his beloved Nell, with the rest of her 'varsity hockey team. Mertz admired these pictures greatly and demanded that when he visited Sydney he should be introduced to Nell and the rest of the team. On birthdays, it was the custom to celebrate the event not only with a special meal, but also by the composition of a comic song about the celebrant. The song about Mertz describes him as 'far too fond of all ze girls' and suggests that he has embarked on the expedition to escape from a romantic entanglement.

There was also a serious side to Mertz, however, which emerges in an article he wrote for the expedition's newspaper, the *Adelie Blizzard*, celebrating the 'danger, cold and hardship' of mountaineering. It is as if instead of exploring the dangers and insecurities of family and sexual relationships, Mertz escapes to the more impersonal challenge of the physical landscape. He writes that the polar explorer is said to 'love the lonely inhospitable ice-regions' as the 'mountaineer loves his mountains above all things'. But such a love can exact a very high price for its exhilarations.

• • •

A pallid sunlight sliced into Mawson's makeshift tent, encouraging him to immediate action. As he prepared to set off, he hoped that the light would last long enough for him to make some real progress. He had already packed most of his gear in anticipation of a move, but still he had to take down his shelter, which entailed moving the ice blocks that held down the canvas, removing the tent from the wooden frame, unknotting the rope that secured the three sawn-down slats of sledge into a tripod, and packing all this onto the sledge.

By the time he had his harness on and his newly fashioned rope ladder slung over his shoulder, Mawson felt giddy and faint. The temptation to lie down again and let the elements destroy him was almost overwhelming. But he was not a quitter, he said to himself; he was, in Service's words, *a fighter from away back*, and put one foot in front of the other.

As he plodded forward he tried to fix his mind like a compass needle on the shattered terrain through which he had to pass. And in a way the very difficulty helped him, for the uneven surface and the pock marks of crevasses meant that he had to focus on the march to avoid falling. But if the going was good even for a few steps he found his mind wandering to the usual subject of food. The craving for sustenance, the gnawing need in his belly, took his mind down tracks of abomination. He found himself wondering whether, after all, he should have butchered Mertz's body for supplementary meat. At the time he had persuaded himself that he had enough rations to get through if the weather was kind, and he had baulked at the idea of performing such a bloody operation on the body of a man who had been his comrade. His determination to cling to the ideas of civilisation, however primitive his situation, had been encouraged by the idea that Mertz's whole body was poisoned anyway. This was the only explanation for his demise; it was clear that the mental and bodily collapse of the tough and stocky Swiss was total.

Mawson managed a thin-lipped smile as he thought of all this and incongruously remembered the celebrations in Sydney when as a young man he had returned from a geological expedition to the New Hebrides. As part of the graduation week festivities he had covered himself with blacking and dressed as a cannibal, stirring a large pot on the back of a float that caused much hilarity as it passed through the streets. The irony seemed very sharp. The unforgiving hostility of Antarctica had humbled his sense of superiority. He knew now that it was possible for all human beings to consider breaking any taboo in order to save themselves. His imagination now, as then, could besmirch itself with blood in the name of hunger and survival. He consoled himself that in the end he had not acted out his impulses. Only time would tell whether this was the right decision or not.

Mawson was jerked from such reflections as his feet sank through the lid of a crevasse, and he felt again the rush of the fall. It was a sensation which did not improve on acquaintance. But on this occasion he had his new ladder. Mawson climbed the flimsy, swinging rungs to safety, breathless but jubilant that his own ingenuity had saved him. After a brief respite he was ready to go on. But the sun had disappeared, leaving him in lowering light with thick drift, and all round him were obstacles to progress. Mawson drove himself a little further, then faint and dizzy with fatigue he surveyed his surroundings. He could no longer be sure of his direction, and under the circumstances it seemed futile to carry on, using up energy for questionable progress in such dangerous conditions.

So he set about making camp again. This was an arduous operation, entailing as it did cutting ice blocks with his spade, knotting together the bits of wood that formed the improvised tent frame with frozen, fumbling fingers, and then spreading the canvas over the top and weighting it with the ice blocks. It was an impossible operation to perform in any sort of wind, but today, thankfully, despite the poor light, it was not too strong. But still it took a monumental struggle, and hours of intense labour before Mawson was in his shelter, tasting a little welcome food, and attending to his diary:

> Glimmer of sun at 10am, so I got out and off but [it] soon disappeared. I sank to knees in one crevasse and got out and, as sun went, noted open ones all around, so decided to camp and consider matters and be ready to start if sun appeared again. Now ration is being reduced blood is going back and several festerings broken out again. If only I had light I could make the Hut. Had I known of all this overcastness to be expected I would have tried for the sea but it is too late now as I am in a maze of crevassed serac ice with no light ... I was quite faint at end of 1/2 m today and know that I can't stand much heavy work.

• • •

When Ninnis was lost down the glacier, Mawson and Mertz were left with six dogs and a week's rations for three men, enough for ten days between two of them or twenty days for one. Mawson also had a few raisins and sticks of dark chocolate. The sledging ration was very like Scott's, and amounted to 34.35 ounces for each man each day. The week's ration for three men came to about 42 pounds. Mawson's diary indicates that between Ninnis's accident and the death of Mertz their primary diet was dog, although this was supplemented by 4–5 ounces a day of pemmican, biscuit and cocoa, a few raisins or some chocolate. When Mertz became ill, they ate less dog and more of the sledging rations. This means that when Mertz died, Mawson would have had in excess of 20 pounds of sledging rations left, and some dog-meat. He had been eating approximately 14–16 ounces a day since the accident, so he probably calculated that if he covered five miles a day for twenty days, he had enough food to do the hundred miles back to the hut.

On 6 January Mawson wrote in his diary, 'I could pull through with the provisions at hand.' It is difficult to believe that a day or two later he had changed his mind. Yet on 8 and 9 January Mawson casts doubt on his ability to survive. But this may not have to do with provisions. After Mertz's death at 2a.m. on 8 January, Mawson spent the 8th, 9th and 10th unable to move because of poor weather. He spent the time cutting the sledge in half, cooking the rest of the dog meat, and otherwise modifying his gear ready to make his dash for safety. He also records building a cairn of ice blocks around Mertz's body and surmounting this with a cross made from the cut-down sledge runners. He read the burial service over Mertz on 9 January. Standing there alone in that great wilderness of ice, intoning the solemn cadences of the burial service, he wondered perhaps as he read who, if anybody, would perform such a service for him.

Mertz's death and burial, the delay due to poor weather, and Mawson's cognisance of how difficult it was going to be to sledge and camp alone, were surely enough to cause a depressive reaction. Once he was travelling again, hope was immediately regained. On 11 January, after his first day's journeying, he says, 'If Providence can give me 20 days weather like this ... surely I can reach succour.'

The full sledging ration of 34 ounces was not enough to keep Scott and his men healthy, and this was also true of Mawson's party. Malnutrition was exacerbated for Mawson and Mertz by eating the livers of Greenland Huskies which are very rich in vitamin A. Eating even relatively small quantities of this results in a clinical condition called hypervitaminosis A, which at first has symptoms such as headaches, vomiting and diarrhoea followed by scaling and stripping of skin, loss of hair and the splitting of skin round the mouth, nose and eyes. In its more advanced stages it causes irritability, loss of appetite, drowsiness, vertigo, dizziness, skeletal pain, loss of weight and internal disruption from swelling of liver and spleen including violent dysentery; in severe cases this leads to convulsions and delirium and possible death from intercranial haemorrhage.

Mawson's account of Mertz's death mentions most of these symptoms. And Mawson himself was also suffering from some of them, though obviously he was less severely affected than Mertz. This may have been because he ate less of the liver than Mertz, that in trying to be kind to his companion he let Mertz have more of the richer meat, thinking that it would be good for him.

At the time, Mawson wrote in his diary that Mertz's death was due to 'exposure finally bringing on a fever, result of weather exposure and want of food'. Later he spoke of the possibility that Mertz had suffered from ulcerative colitis, and then came to the conclusion that his demise was due to peritonitis. Given his scientific background, Mawson would have been very wary of eating Mertz's body since he described it and his own as 'rotting'.

Mawson battled on, horrified that he had even contemplated eating Mertz, and wondering, as his own plight grew worse, if he had done the right thing.

19 January, 1913

MAWSON SAT AT A round table covered with a white linen tablecloth and set with silver cutlery and cut-glass wine goblets. In the centre, a red rosebud peeped from a delicate crystal vase. The lighting, although from chandeliers, was subdued, and with the deep red of the plush wallpaper conveyed an atmosphere of enfolding warmth. Ninnis and Mertz, dressed in dinner suits, discussed their preferred desserts, while Mawson gave thought to the roast beef that was to come.

A waiter poured a rich plum-coloured claret into their glasses. And then a great domed salver arrived, and its lid was removed to reveal the joint. It looked smaller than he had anticipated, but even so Mawson's mouth began to water as the waiter began to carve and a plate of meat was put before him. As he lifted his knife and fork, he realised that the meat wasn't cooked properly: it oozed bright red blood and neither smelt nor looked like beef. Someone screamed, 'You can't eat that!', the plate was whisked from beneath his nose, and Ninnis and Mertz disappeared. People dressed in high fashion were pointing and jeering at him as if he had done something shameful. He wanted to run but was trapped by their taunts, which grew louder and louder, voices booming and echoing remonstrance.

Mawson woke confused and disoriented as he felt spittle in his mouth from its famished anticipation and heard loud cracks and cannonades filling his consciousness. He disentangled himself from his dream as he scrambled out of his sleeping bag. He sensed thin swathes of light penetrating the gloom of his tent and, sure enough, as he untied the cords that bound the canvas covering of his shelter, he emerged to find a pallid sun inhabiting a clear sky. He looked at his watch. It was 4a.m. The glacier continued to creak and groan and boom as Mawson straightened, feeling the gusts of a strong wind tug at him. As he clumsily prepared to urinate, he had to brace himself back to the frozen air, and his mind filled with the eeriness of the noise.

The glacier was alive with opposing forces, as if nature was at war with itself. The crashing din made Mawson feel smaller and more alone than ever. He rehearsed what he took to be the scientific explanation: that the sounds must be the result of re-freezing and splitting of the ice as a result of temperature changes. But the loudness of the explosive reports seemed to escape reason and embody more primitive impulses. It was impossible to dispel the idea of some titanic underground battle to which Mawson was the flea-like observer.

It was with some sense of relief that Mawson regained his shelter and, closing out the elements, huddled back into his bag. The wind was far too strong to risk moving; he would have to wait for better things later in the day. For now he curled up, shivering, holding himself together by talking to his God. The magnitude of the forces ranged against him moved Mawson into a passionate self-persuading colloquy in which he invoked an all-powerful presence to give him strength to reduce the problem of his survival to a matter of reason and will. The riven landscape outside his tent seemed like a labyrinthine trap; a place of death and sterility imbued with mysterious power. Mawson knew that if he were not to go mad he had to control such perceptions and conquer the brutal and inanimate; to see with the help of God that he too could muster even greater power and keep going to the hut.

Or was all this a punishment for some unknown transgression? Was he fooling himself with the sense of a benign presence helping him? Why should he survive after Ninnis and Mertz had perished? Surely no God of love would have saved him from the crevasse only to torture him further before death?

Love. Mawson thought about this strange word and what it meant. He was not often given to such reflection. Soul-searching complexities he considered a waste of time. Human beings were either good at something or they were not; useful or useless. Mawson's parents had brought him up to be useful, to be a professional. They had been rather distant physically and emotionally; they were not gushers but emphasised discipline, achievement and hard work. Action not words, Mawson realised, was a motto he had adopted in part at least from his parents' teachings. He had also inherited his short temper from them, which he recognised to be one of his worst faults. And it was from his parents' unpredictable temper that he had first learned to turn away from human uncertainties to find his vocation in the study of science with its wonderful laws, impersonal beauty, and abstract reason.

Towards Mertz and Ninnis, Mawson could not say that he felt love. He hoped, however, that he had acted decently towards them as they had towards him. They were both great doers, cheerful workers, tough characters. This was what he admired. They had been the logical choice to accompany him as they had looked after the dogs through the journey to Antarctica and throughout the winter. But he had also vicariously enjoyed the closeness between his two companions—they provided an image of intimate friendship that Mawson knew he would never experience himself. He was not, he thought regretfully, that kind of person. He liked to remain a little aloof; that way no one could take advantage of you.

But he dared not think of Ninnis and Mertz for too long now, for then his sense of aloneness threatened him. It was no good looking backwards for love, he must concentrate on the future. And in the future was the possibility of Paquita. Mawson thought of her bright intelligence, her good humour, her kindness. She was so unlike him, with a lively outward-going personality he found irresistible. He was also sure that she was upright and virtuous. And then, Mawson knew, his feelings for Paquita were also about the mysteries of sex. She had an aura that intoxicated him and filled him with longing.

Actions not words, Mawson repeated to himself. His actions would show his love for Paquita. He would walk back alive to save her from grief and to claim his place beside her. He would live, and he would ask his God, his Providence, to show love in preserving him from the cruelties and dangers through which he must march. With such thoughts in mind, Mawson allowed himself to drift into a light sleep, feeling sure that later the day would allow him to progress.

• • •

Mawson stood in the midst of a ruined landscape gulping air in harsh gasps. All around him the glacier ice was crushed and buckled into fantastic formations, and everywhere were striations marking crevasses, many of which were filled with snow. He had set out at 8.30a.m. and been forced to zig-zag through the creased and corrugated labyrinth, every moment expecting to be plummeted into the depths. Now, dizzy with fatigue and seeing only more tribulation before him, his spirits sank.

The sight of so many crevasses inevitably turned his mind to the loss of Ninnis. On that day they had all been feeling elated. They had battled over the glaciers and were travelling well across a flat expanse of snow, knowing that in a day or two they would turn their faces towards home. They had rationalised their gear onto two sledges and Mertz was ahead on his skis, playing the role of pathfinder, and yodelling his student songs into the scimitar air. Mawson remembered the scene with a disturbing vividness. The sky quintessential blue, the white purity of the snow, both colours he now knew could speak of infinite pleasure or infinite pain. He had crossed a crevasse obliquely and turned to call back to Ninnis to watch out for it. Then he had carried on, not thinking it was particularly dangerous. After a little while he saw Mertz turn and signal inquiringly back to him. Mawson turned and saw nothing but the snowfield glittering to the horizon. Ninnis, the dog-team and sledge had simply disappeared.

Mawson felt sick as he recalled the terrible events that followed. He and Mertz had found the gaping chasm eleven feet across and seemingly bottomless. They had called and shouted for hours, but with no response. Mertz had been overcome with grief, weeping unashamedly and desperately. Lying flat on their stomachs, they peered over the ledge and could just discern a dog whining, its back broken and stuck on a ledge many feet below. Of Ninnis there was no sign at all. They let down a fishing line and found that the ledge was one hundred and fifty feet below. All the rope they had knotted together would not reach so far. Nevertheless, they took it in turns to hang over the edge of the crevasse supported by alpine rope in order to yell and peer into the darkness.

There was no response, only a clearer view of one dead and one dying dog lying next to what appeared to be the tent and a food tank containing enough provisions for three men for a fortnight. Grief and frustration and fear had combined into a terrible conflict of emotion. Mertz's expression of his feelings made it easier for Mawson to restrain his. He had to play the role of leader, relieving his distress by decisive action. He insisted on taking exact bearings of where they were, supervised the arrangement of their depleted stores and equipment, and finally read the Anglican burial service by the cavernous ice-grave. He and Mertz had stood together, bare-headed in the perishing cold, as Mawson intoned the solemn cadences designed to comfort and console, but which on this occasion reminded them more of the reality of sudden loss and the fragility of their own mortality.

Now as he stared at the maze of crevasses around him, Mawson felt that he too would soon be joining Ninnis and Mertz locked in the frozen embrace of this hostile land. But to be still for long could not be borne, and inaction was a misery to him. The only option was to move forward, making his way as best he could, and pledging his faith in the Providence that had guided him so far. *Yea though I walk through the valley of the shadow of death, I will fear no evil: for thou art with me; thy rod and thy staff they comfort me.* The words came into Mawson's mind unbidden, and he thought of all the Sundays when he had led prayers and hymn singing back at the hut. That he should remember these words at this moment seemed like a sign to him. He must go forward and nurture hope. At least if he survived, he could tell the story of his lost companions, rather than condemn their names to being lost to this desolate silence.

• • •

Although it is true that Belgrave Ninnis's name has survived, he remains an even shadowier figure than Mertz. He was only twenty-three when he died, a 'fair complexioned giant' who stood at six foot, four inches and was an English lieutenant in the Royal Fusiliers, one of the swagger regiments. In two portrait photographs of him in full dress uniform he looks suitably solemn, and his rather thick underlip and bowed top lip give him a slightly sulky demeanour, suggesting the pessimism that one of his fellow-expeditioners identified as part of his character. But with his fresh-faced English complexion and round, ruddy cheeks it is easy to see how he earned the nickname 'Cherub'. Another photograph shows him sitting on the rail of the *Aurora* in collarless shirt, braces and striped trousers. He looks straight at the camera with a cheeky grin and ears that stick out too far, his hair is centre-parted, his feet are too big: he looks painfully young and painfully thin. This is perhaps the Ninnis to whom Mertz adopted a maternal attitude. And like Mertz, Ninnis was much admired by his fellow expeditioners, he was willing to do any of the dirty work and was frightfully keen on the expedition.

Ninnis had a good pedigree for exploring. His father had served as a naturalist on the South Australian surveying expedition to the Northern Territory of 1864; ten years later as a Surgeon Captain in the Royal Navy he had served with the Arctic expeditions of the *Alert* and *Discovery*. He had evidently passed on his enthusiasm to his son, who had unsuccessfully volunteered for Scott's expedition before being accepted by Mawson.

There is nothing to suggest that Ninnis's childhood and education were anything other than typical of his time and class, imbuing him with those typical Edwardian values which located manhood in adventure, pain and self-sacrifice. 'Gloomy' Davis remarked of Ninnis that he had the 'manner and character ... of an enthusiastic public school boy', and that while he was not as agile as Mertz, he 'had made himself into a competent seaman' during the voyage from Hobart to London. Despite at least one of the Australian expeditioner's reference to the fact that Ninnis had 'more boxes of beautiful clothes than seemed possible for one mere man', social and national distinctions do not seem to have impinged on his relationships.

On the march with Mawson and Mertz before his death, Ninnis seems to have suffered physically more than the other two. The conditions through which they pulled and steered their sledges and dog-teams were appalling. Ninnis was the biggest of the men, but they were all on equal rations. It was therefore Ninnis who first began to show signs of physical deterioration.

Only a few days after they'd embarked on the journey they were forced by atrocious weather to remain in their sleeping bags for two days. Mertz read Sherlock Holmes aloud to them, practising his English. No progress meant that they immediately adjusted to reduced rations, with Mawson reporting that they all felt 'pretty rotten' due to the inexercise but making special mention of Ninnis, who was 'quite faint at noon'.

Three weeks later on 6, 7, and 8 December, they were laid up again and we first hear mention of Ninnis's finger being infected with a whitlow, an inflammation of the deep tissues that causes swelling and a severe throbbing pain. Ninnis's sleep was disturbed by this and by acute dreams. Mawson records him sitting up in his sleeping bag crying 'Hike, Hike' to the sledge-dogs who were pulling him through his agony.

For another five days Ninnis endured increasing pain without complaint. Mawson noticed that he sat up at night for hours clutching the comfort of pipe and book, unable to sleep because of the acute discomfort of his swollen hand. For all that, Ninnis continued to do his share of the work without grumbling, until on the morning of 13 December, after a very bad night, he asked Mawson to relieve his suffering by lancing the finger. Mawson sterilised a knife in the primus flame and Mertz cradled his friend's head. Then Mawson, having first chilled the finger with ice, holding his breath, pressured the blade of the knife down the side of the swollen and discoloured digit. Ninnis groaned through gritted teeth. Mawson administered some water laced with a little alcohol, and the patient was wrapped in his sleeping bag to recover from his ordeal. The pain subsided and the trio continued their journey that afternoon.

The following day Ninnis lost his life in the crevasse. Mertz crossed the crevasse on skis. Mawson rode over it on the sledge. Ninnis, however, attempted to walk across it beside his sledge. Mawson later thought this was the crucial factor: Ninnis's body weight was not spread but concentrated, so he went through the snow crust and the dogs and sledge followed.

. . .

Mawson sipped a weak mixture of alcohol and hot water in the comparative comfort of his flimsy shelter. He was both celebrating and preparing himself for another ordeal. That afternoon, after nerving himself to continue his march, and after several times breaking through the crust of crevasses only to be saved by his arms or by his new rope ladder, he had seen the land open out before him and known that at last he had crossed the glacier. He had travelled a little further, to the base of the slopes that led to the Adélie Land plateau three thousand feet above his present position. But the light and surface had deteriorated, so now he was camped preparing to begin his climb up the snow slopes tomorrow.

It had been a good day, and he was at last across the glacier. He was filled with renewed hope, and determined to spend his time productively. His first task, however, was not a pleasant one. He had a boil on his leg that was giving him discomfort, and he had resolved to lance it. The alcohol was in part a preparation for this operation.

Mawson decided that while he was attending to his leg, he may as well treat the rest of his body, anointing the raw and scabbed parts of his anatomy with lanolin. He sipped the warming alcohol as he began to undress. It was both awkward and excruciatingly cold. Still Mawson persevered; it was, after all, his body that would get him home. His feet, legs, the tops of his thighs and his scrotum were in very poor condition. Where the top skin had peeled away, a blistered layer of thin, watery substance had formed as new skin attempted, but failed, to grow. It was unpleasant to contemplate, yet Mawson knew that he must do what he could to alleviate it. He smeared himself with lanolin, taking great deep breaths as he stung the raw parts of himself with rubbing fingers.

Then there was the boil. Mawson followed the same procedure that he had used with Ninnis, but there was nobody to hold him while the knife did its work. He gritted his teeth. He would imagine holding Paquita on the verandah in Adelaide with the sound of the sea caressing Brighton Beach in the bosky light of a summer's evening. He sterilised the knife and, concentrating on the picture in his mind, lanced the boil. For a bleak, shuddering moment pain governed all; mind and body were all red agony. Then, as he bound the wound with iodine, he sipped more spirit and began to struggle back into his clothes.

The action warmed him and the intensity of the pain began to subside. He lay for a while recovering, thinking of the clarity of his Paquita fantasy. It was not so much that he could conjure her face in his waking dreams, but that he could bring her mysterious presence to life in his mind in a way that he never had before with any other person.

But it would not do to dwell too much in the realms of romance. Memory and imagination had to give way to the practical tasks that claimed his attention once more. 'Fight! That's right. I must struggle. To rest means death,' he told himself. Part of this struggle, he recognised, was to concentrate his mind on the task at hand. Mawson turned to the job of waxing and tarring the sledge runners. He sat talking to himself and his wooden companion: 'We have to look after ourselves; clean ourselves up, be efficient, that's the way to go. Strength, will, determination and Providence.' He repeated the words like a prayer cut in stone.

The sledge prepared, Mawson then went over all his gear, selecting what could be discarded to make his ascent to the plateau easier. He decided that some alpine rope, worn crampons and socks could be left behind. Lastly, before turning in to sleep, Mawson wrote in his diary:

> Everything seemed hopeless—the serac seemed to be endless, the glacier cracked and boomed below. It seemed impossible for me, alone, to cross it, for

any moment I expected to go down. I determined at last to stick to the course as much as possible, push on and rely on Providence.

After 2 miles I was surprised to find the landscape opening out, and it appeared only several miles to the Adélie Land slopes; also seracs disappearing . . .

Later, as snow got too wet to travel, I decided to turn in early and go on in early morning hoping for a good day. So I did not have the 'tea'. Want light to go ahead, as much crevassed still. Have great hopes for tomorrow. I took off all clothes to get at boil on leg and stuck it. Had general clean up—polished sledge runners and tarred them, went over gear and discarded some that not essential.

20 January, 1913

MAWSON STAGGERED FROM HIS TENT for the second time that morning. It was 6a.m. At 3a.m. he had been outside to pump-ship and stood watching a glimmer of sunlight fade into a symphony of grey. To the west, he could see the snow slopes ascending to a craggy peak that seemed to loom over him threateningly. Behind him were the riven icy walls of the glacier. To his dismay he could just discern from where he stood, the lines of crevasses, making his passage onwards as hazardous as ever. He needed more light in order to travel safely, so he had retired to his bag, hoping for an improvement in conditions later on.

Now he was disappointed. Sky and ice, mountains and snow-fields were all indistinguishable, reduced to a leaden curtain, blank and impenetrable. Mawson took the broken and much-repaired shovel and dug a shallow hole in the soft snow. As he dropped his trousers to perform his 'rear', he remembered the incomparable mornings in the Flinders Ranges when, perched above some dry creek bed, he had accomplished the same function serenaded by the rich syllables of warbling magpies. He thought of the vaulting of the outstretched branches of red and river gums, the pillars of their trunks on either side of the stony aisle; nature's great cathedral where he had, he now considered, worshipped without knowing that that was what he was doing.

The contrast in his circumstances was made more ironic by the knowledge that it was his attraction to the pre-Cambrian rocks of the Flinders Ranges that had inspired his interest in Antarctica. The mountains of South Australia had evolved from the Ice Age into their brown, green and grey beauty, and were now decorated by the magnificent flight of the wedge-tail eagles that haunt the thermals there; and by the wonder of the fabulously varied eucalypts through which, if you were lucky, you might see the delicate face of a rock wallaby peering, transfixed at intruders into its territory.

Mawson gathered his clothes about him again and shivered. If the barren landscape that pressed in upon him was appalling in its deathly power, he found satisfaction in the thought that his body was still functioning. He had always marvelled at the efficiency of the human body in its digestive functions, and enjoyed the sense after his morning toilet that he was ready again to partake of the day's nourishments relieved of waste. Although such reflections belonged to happier times and despite his thin and pitiful stool, he took comfort from such thoughts back to his tent, and began to prepare a scant breakfast.

Mawson only rewarded himself with pemmican hoosh when he had either travelled or was confident that he was about to make some ground. This morning no such certainty existed, so instead he took some of the remaining jelly that he had made by boiling down dog paws and mixed this with some hot water to make a thin broth. As ever, the heat from the primus melted the frozen condensation of his breath on the inside of the canvas, and this began to drip on him. It was miserable that the otherwise benign presence of heat and light from the stove was also accompanied by this discomfort.

As he sipped the tasteless brew, craving strong flavour, Mawson was reminded of Ninnis's famous cooking 'championship'. Cherub had been attempting Mrs Beeton's recipe for kedgeree but had confused the final instruction, which resulted in him putting eight ounces of pepper and eight ounces of salt to season four large tins of salmon. The dish had looked perfect coming from the oven, but there was much choking, perspiration and ribald humour when people tasted it. Any such mistakes in the round of domestic or scientific chores were deemed 'championships'. Ninnis's was the greatest of them all.

Mawson remembered the good-humoured chaffing, the warmth and bustle of the hut, the hard work during the day and the sing-songs in the evening. It was a little space of civilisation in the vast and hostile wilderness. He wondered if he would ever experience it again or taste real food. To banish these bleak thoughts Mawson decided, on finishing his poor repast, to pack his sledge and make ready to start as soon as possible. But without taking down his shelter this was only half an hour's activity, and soon he was crouched in the tent again, praying for a change in the weather. Through a long morning Mawson watched and waited, trying to keep panic at bay by focusing on the task ahead, and summoning what patience he could to watch the stony weather for signs of change.

• • •

Mawson's diary entries show his belief in the value of physical activity. For him, being unable to continue the trek must have been excruciating. Alone, under enormous physical and mental stress, in constant fear for his life, to sit and do nothing without collapsing mentally must have taken huge reserves of will; the ego huge and unwavering, a rock resisting the powerful instinctual responses that threatened it with instability and dissolution. Any thought of the hut and his comrades must have been like an ambivalent dream, tantalising with promise yet reminding him continually of his solitude.

For though there had been tensions in the hut through the long and arduous polar night of winter, there had also been festivity and celebration. Mawson had maintained his distance and his role as leader by providing an example both as a hard worker and someone who could join in the fun and games on occasion. He allowed the men (whom he often refers to as 'the staff') a liberal amount of chocolate, alcohol and tobacco. Birthdays, high days and holidays were named 'suspicious occasions' and celebrated in style. When none was in the offing the almanac was consulted and a suitable party contrived. This resulted on one occasion in a fine dinner to celebrate the anniversary of the first lighting of gas lamps in London.

There were also practical jokes. Hurley, the photographer, was the leading comedian, but he was not short of willing collaborators. So the nervous and fussy John Close was teased about the propensity for acetylene to explode and terrorised one day by the others who set off small acetylene bombs around him. Hurley took a rise out of Archie McLean by putting a dead penguin into his bed just before he got up. One morning there was a rough and tumble between the 'East-enders' and 'West-enders' which ended with pillows, blankets and bodies flying in all directions.

To fill in the slow evening hours there was no shortage of amusing raconteurs. Murphy and Hurley particularly were a fund of tall tales, the latter claiming he had been jilted twenty-six times and not being shy to tell his comrades all the details of these affairs. Several of the others also shared stories of their romantic entanglements or disappointments. On his birthday, McLean woke up to a mass of presents including many pictures of girls culled from magazines and suitably inscribed.

Reading the diaries of Scott and his men, it is easy to gain the impression that they were more interested in each other than they were in women. This is in stark contrast to the Mawson expedition, where there seems to have been considerable badinage about the opposite sex.

Sometimes there was a more serious side to the 'suspicious' occasions. On his birthday, for instance, Mawson gave a speech on the stages of a man's life which Hunter, for one, enjoyed, but concluded that Mawson was 'too much of a materialist'. At another dinner, this time to celebrate the 'annexation' of Cape Denison in the name of King George, Mawson took the opportunity to give the men a pep-talk, telling them that they were in for a hard time and that they needed to put their 'shoulder to the wheel in the coming fight and uphold the honour of Australia and the British Empire, especially as a German and two French expeditions had refused to land in these regions, declaring them to be too difficult for carrying out any work'.

Though a photograph of this occasion shows the hut festooned with Union Jacks, Mawson sat at the head of the table wrapped in an Australian flag.

Mawson was being forthright in his speech. He soon realised that he had situated the hut in one of the most inhospitable areas of Antarctica with extreme weather a constant threat; gale to hurricane force wind and darkness prevailing beyond the confines of the hut for all of the winter. From March to October the average wind velocity was never less than strong gale force on the Beaufort scale. Nothing like it had previously been experienced or recorded anywhere else in the world. For a fortnight in May the average did not fall below seventy miles per hour—between storm and hurricane force. On 14 May the average for twenty-four hours was ninety miles per hour. Such averages mean gusts of much greater strength travelling at well over one hundred miles per hour—a gale strong enough to level cities.

The winds increased the stress of the winter months. It meant that the expeditioners were confined to quarters for long periods at a time under artificial light, and there were times when they thought that their roof was about to be blown away. It is a tribute to the design and placement of the hut that it survived at all. It was well built, but it was also situated below the rise where the anemometer was measuring wind in a small shelter built specifically for that purpose. The living quarters were bolstered and protected by drift snow that virtually buried the hut. At the height of winter, egress was gained through a series of snow tunnels or a trap door in the roof. And although this was inconvenient, it meant that the structure had further protection from the blast.

Mawson and his men took their turns to brave the elements in order to read the scientific instruments, often a painful, not to say heroic, endeavour given the conditions. The wind was often strong enough to knock men off their feet and bowl them along. Crampons were essential, but even with these, 'hurricane-walking' was an acquired art in which men were often reduced to crawling over the ice on their hands and knees.

Mawson believed that work and constant activity were the only antidotes to the huge potential for depressive effects arising from their circumstances. Scott's expedition ran on a model of navy discipline and class-distinction which tended to formalise divisions between its members, but Mawson's organisation was civilian, scientific and more egalitarian. Scott's hut was divided in two by a wall of packing cases which separated the officers and scientists from the other ranks. Cooking and cleaning duties were undertaken only by the latter. No such barriers existed in Mawson's hut, where each man was rostered to take his turn as cook, messman and night-watchman. Apart from these ubiquitous duties, most of Mawson's men also had a specialised mechanical or scientific job. Of the few non-scientists, Ninnis and Mertz were in charge of the dogs, Hurley was the photographer, and H. D. Murphy, sometime scholar of history at Oxford University, was in charge of stores. Despite their various backgrounds, only Ninnis, Close, and Hurley had no experience of tertiary education. There was a more professional atmosphere to Mawson's expedition than Scott's, a more unified sense of purpose.

For all that, in the course of the winter, various members of the expedition managed to annoy Mawson.

His approach to the foibles of human character and the intricacies of personal psychology are as unsophisticated as they are ruthlessly practical. In his diary he records his conception of the four classes of men: those who are accomplished and painstaking stickers; those who are not really good at anything but can assist under supervision; those who require winding up; and those who don't fit in and are no good at anything. Happily for all concerned most members of the expedition fell into the first category. But others were upbraided when they failed to meet Mawson's standards. Webb is chastised for reading at table, and Stillwell has to be dealt with in a row over the work roster. Mawson writes: 'He continued to sulk so taunted him to develop a row but he was very feeble. His chief fault is that he is so slow in getting a move on.'

Mawson very rarely records praise of any of his colleagues, but criticism is given strong voice. And this is particularly concentrated upon three men: Close, Murphy and Whetter. These three had the least to do with the scientific and sledging program. Close's job is vaguely described as 'assistant collector', Murphy was the storeman, and Whetter a medical doctor. John Close, at forty, was the oldest of the party at Cape Denison. He was a veteran of the Boer War and, according to Hunter, 'would have made a good servant'. He became the butt of many jokes as he was forever fussing over and repeating all of the 'Doctor's [Mawson's] orders'. He gives the impression of being a rather pathetic figure, bumbling and sycophantic.

Herbert Dyce Murphy was a charming and gentlemanly figure who had read history at Oxford but left without taking his degree. The other men liked him for his ability to tell stories. Latterly he has gained some measure of fame or infamy as the transvestite who reputedly provided Patrick White for the model of his character Edie/Eddie in *The Twyborn Affair*. Whetter is a less colourful character who tended to be lazy and to object to the kind of manual work that Mawson demanded of him.

Lack of effort or perceived weakness form the basis of many of Mawson's complaints. Close first attracts Mawson's ire by retiring to bed with a 'severe gum boil'. Rather than arousing sympathy, this draws forth the comment, 'He should never have thought of coming to the Antarctic with such remnants of teeth.' Going to bed early is also the source of further trouble a few weeks later:

> I have asked Whetter to put in all the time he can preparing clothes for sledging. He has taken no notice and retires to his bunk at 4.30pm as usual.
>
> Close has put in much time at his clothes, but gets tired before the day is out and has a nap at intervals. After 4pm he is prone to read a book and does the same when he is not asleep till after midnight.
>
> Murphy does some sewing but very feeble. After dinner he never does anything but go to bed and read, or (as of late) design collages.
>
> Neither [sic] of these 3 wonderful people have contributed to the Blizzard or attempted to.

Mawson clearly believes that these three are being lazy, and that reading is an activity to be pursued only when all other tasks have been fulfilled; an activity of last resort.

As the winter went on, complaints about Whetter retiring early continued, and Murphy came in for further reprimand: 'Murphy was engaged spoiling his boots today in the belief that he was mending them. I had occasion to speak to him.' Whetter, on having diarrhoea and complaining the following day of not feeling 'too well', attracted Mawson's judgment that the offender was 'not fit for a polar expedition'. He is thoroughly annoyed when Whetter asks to be excused his duties as nightwatchman. Two days later Mawson writes, 'Whetter cooks but makes a hopeless failure of it.'

Later in the winter there comes this incident:

> At something to 4pm Whetter came in, took his clothes off and intended to read a book. Before lunch I had asked him to dig out the hangar in front after getting in the ice.
>
> I heard at 4pm that he had not done this, and his appearance in the hut to read a book was in direct disobedience of my orders. I was very wroth about this and asked him why he was coming in under the circumstances. He said he had done enough. I asked him what he had come on the expedition for. He said 'not to do such kind of work', I said he was a 'bloody fool to come on the expedition if that was the case.' He said 'Bloody fool yourself' and 'I won't be caught on another one.' I instantly told him to come into my room. I was wild but immediately calmed and talked things over with him at length in the most lenient and persuasive terms possible to try and let him see his error.

Whetter was unconvinced, and the result of the row was that Mawson gave an address to the assembled company at dinner about the need for the 'united efforts of all . . . to make the expedition successful'.

Mawson's work ethic, and his definition of 'work' as necessarily physical, is offended by these men. He is also convinced of his own rightness; he is the leader, he should be obeyed. However unpleasant his aggression, and the tone of superiority in the diary, here was a man who needed to assert himself and who did not shirk confrontation when he thought it necessary.

• • •

Mawson bent at the waist and heaved his sledge forward. The light was closing in again, and the wind was strengthening against him. All his hopes were gradually being extinguished by the cruel and chameleon weather. He had waited until 2p.m. before in desperation he had set off, feeling that he had to make some progress at whatever risk. At first the going had been good, though the pulling was hard uphill. But now with conditions worsening, Mawson felt a creeping blackness invade his mind.

He thought of all the battles he'd fought to make the expedition possible—the wrangles with Scott and Shackleton, the desperate lobbying for money and support in both Australia and England. That all these efforts should end in ignominious defeat seemed terrible beyond words.

But he would not give in to the demon of despair and surrender. He brought to mind the lines of Service from his poem the 'Trail of '98':

> We tightened our girths and our pack straps; we linked on the Human chain
> Struggling up to the summit, where every step was a pain.
> Gone was the joy of our faces, grim and haggard and pale;
> The heedless mirth of the shipboard was changed to the care of the trail.

Although he was alone, Mawson determined to keep proving himself. And the sheer fact of who he was kept him going. He came of tough Yorkshire stock; Australia had made him tougher. Other men may weaken and fail, but he had never shirked or wavered from whatever he wished to achieve. He would show Scott and Shackleton what he was made of, as he had shown them before. He was hard, strong, resilient. He was iron. He was one of the 'silent men who do things', and nothing would erode his self-belief.

Mawson considered in his hardship that those who spouted about the value of suffering in sentimental ways were fools. Suffering was a necessary part and process of life. If you had to suffer, then you bore it with stoicism and with pride. Though he was bowed to the waist as he hauled, in his mind he fought off the assailants of doubt and pain with the story of his own self-conquest.

So he went on, knowing that soon the wind would mean he would have to camp. But not yet. And there was always tomorrow.

・・・

Mawson went to Yorkshire in January 1910 when taking a break from his negotiations with Scott and Shackleton. It was a tense and difficult time. Failure to agree with Scott put Mawson back into an uneasy relationship with that eloquent and charming rogue Shackleton. Mawson had been using Shackleton's Regent Street office and had also spoken to his erstwhile leader about the Cape Adare plans. So he was subsequently taken aback when Shackleton came into the office one morning and said: 'I've decided to go to the coast west of Cape Adare and you are to be chief scientist. I can get hold of £70,000—more than enough to finance an expedition.' Mawson decided to go along with Shackleton because of the money.

Mawson and Shackleton made a good start when they visited a businessman named Lysaght in Plymouth. Mawson talked to him for five hours on the value of Antarctic expeditions and the following day Shackleton asked for Lysaght's financial support. The businessman pledged £10,000.

Shackleton was a man of many schemes. One such concerned the purchase of Hungarian gold mines, the revenue from which might contribute to the £70,000. Never one to miss an opportunity, Shackleton enlisted Mawson to visit some of these mines to assess their worth and profitability. Mawson agreed, excited at the prospect of gold. He set out in March 1910 to Nagybanya, via the Orient Express, and found on arrival 'a small, more or less industrial town set among the low, fir-clad foothills of the Carpathians'. At the same time, Shackleton went on a lecture tour of Canada and the United States. By 15 April, Mawson was in Berlin writing optimistically to Shackleton about the prospects for leasing or buying mines at Beopatak near Nagybanya.

But this letter went unanswered. Mawson, now increasingly anxious to know Shackleton's intentions, decided to effect a meeting in the United States on his way back to Australia. They met in Omaha, Nebraska, where, after a long discussion, Mawson dictated an agreement which he asked Shackleton to sign on 16 May. At this stage, Mawson was beginning to distrust his unreliable colleague. Most of this agreement is about finance except for a clause at the end of the document which said that if Shackleton did not lead another expedition south then Mawson would be commander, although Shackleton would still use his influence in raising the necessary funds. Shackleton might be slippery, but he was evidently no match for the forthright Mawson when it came to such negotiations.

Mawson returned to Australia and his duties at the university. He heard nothing from Shackleton about either the expedition or the mines. With mounting anxiety Mawson eventually presumed that Shackleton had given up the expedition. But then Mawson heard that Shackleton had announced to the Royal Geographical Society precisely the original proposition that Mawson had put to Shackleton. Mawson felt confused and betrayed. Worse was to follow. He cabled Shackleton, asking if Lysaght's money still stood. He received a reply in the affirmative. It was to be several months before Mawson learnt that Shackleton had taken the money himself and invested it elsewhere.

Mawson continued to cable Shackleton seeking a clarification of the situation and in December at last elicited a response: Shackleton would not go on the expedition. Mawson now had to act fast. He decided to make a proposal to the Australasian Association for the Advancement of Science at their meeting on 7 January 1911. He also managed to get several donations from prominent individuals before he left for England again on 27 January to do more fundraising and organisation.

Mawson was miserable in London and, like Scott, hated the business of going cap-in-hand to people for cash. But he was absolutely determined and said that he would go to the Antarctic by whale boat if necessary. Despite being betrayed and double-crossed, Mawson did not break entirely with Shackleton. Shackleton's charm may have been a factor in this, but it is more likely that Mawson's pragmatic side was at work. He was at this stage under massive pressure, having committed himself to an expedition which was to leave in a few months time, and having hardly any guaranteed finance to underwrite it. He still desperately needed Shackleton's help to raise money. Shackleton agreed to help and introduced Mawson to the newspaper magnate, Lord Northcliffe, who agreed to publish an appeal for funds to the public above Shackleton's name in the *Daily Mail*.

Although Mawson had given Kathleen Scott a month's notice that he was going to attempt such an appeal to the public so that she could organise further funds to help her husband, still Scott's supporters objected to the *Daily Mail*'s two-day catch cry of £6000 WANTED TODAY. Scott's patrons, Sir Clements Markham and Sir Edgar Speyer, wrote that Scott, 'The great explorer', still required money, and that his needs ought to be considered first rather than those of anybody else intending to divert support. That Mawson was both Australian and associated with Shackleton doubtless contributed to this response. But Lord Northcliffe's power and influence went beyond such huffing and puffing. In addition to printing the appeal, the editor of the *Daily Mail* sent telegrams to numerous wealthy men in England and Australia and within two days the requisite £12,000 had been pledged. Mawson was shrewd not to pick a fight with Shackleton. However untrustworthy he had proved, he nevertheless was instrumental in finding Mawson the money he needed.

There were further tensions between Mawson and the Scott camp. In March 1911, news of Scott's expedition reached London and it became known that Scott had landed his second party not at the Bay of Whales as was first intended but at Cape Adare. This was the very place that Scott had refused to take Mawson to and now the area of Mawson's proposed base camp. Mawson was not pleased. It was particularly galling in view of Mawson's letters to Scott which set out his own plans in great detail.

Mawson felt betrayed by Scott as well as Shackleton. Cape Adare was no longer a viable proposition for his own expedition. Interviewed by the press, Mawson expressed his dissatisfaction and also despatched an indignant note to Kathleen. She had previously invited him to stay with her in London, but now she was subject to Mawson's irritation. He accused Scott of not being frank with him. In reply, Kathleen invited Mawson to lunch and unleashed her energies upon him. By charm and flattery she brought him round, persuading him that Scott had acted in good faith and that the presence of Amundsen at the Bay of Whales together with the state of the ice had left Scott with no choice. Placated, Mawson did not leave until four o'clock.

Lunch with Kathleen turned out to be one of the more pleasant occasions in an otherwise very trying period. But Mawson's determination was absolute. The landing of Scott's second party at Cape Adare simply meant that he would have to adjust his plans. In both the fundraising and the planning for the Australasian Antarctic Expedition, Mawson showed some of the same qualities that kept him going on his lone march: resolve, willpower, adaptability and pride.

21 January, 1913

ENVELOPED IN DARKNESS, Mawson lay in his bag thinking about food. When he was inactive the savage craving for sustenance was at its worst. There was, he considered, a distinct pattern to hunger pains which included periods when the gnawing emptiness in the gut was so intense as to dominate all thought to the point of obsession; a feeling close to panic as if you could eat anything—boots, bones, or finnesko—just as long as the need was fulfilled. But Mawson had taught himself to recognise this feeling and to know that it would pass eventually, both mind and body somehow accepting that they must go on without immediate nourishment.

So Mawson fought to control his impulse to get up and eat more hoosh. And as he did so memories of other times of plenty came to his mind. Those wonderful days camping in the bush with his brother Will when they had shot rabbits and cooked them over the campfire. The process of skinning and gutting the rabbits came back to him with photographic clarity, and his mouth began to water at the thought of the stew they'd made with spuds and carrots and turnips; the taste of gravy mopped up with fresh damper.

His mind moved to later excursions into the bush, on the dusty track to Arkaroola with the camels, and then at night there would be the feeling of infinite peace, with belly full, lying under the huge scatter of stars which in that country illuminate the cloudless night to such an extent that it is a perpetual twilight. Before his first trip to Antarctica this is what adventure had meant to him. The long treks into the bush doing the geological research that combined intellectual curiosity with the ever-present excitement of the possibility of gaining wealth from mineral finds. He had hoped for so much from the uranium deposits he had identified in the ranges.

Such thoughts prompted a grim smile from broken and blistered lips. Though he was carrying geological specimens with him now, the idea that they would bring him wealth was faintly laughable. Even if he survived, the terrible difficulties of the terrain he had explored made the commercial exploitation of any finds seem next to impossible. And thoughts of those halcyon days when he had felt life opening out before him, expansive as the great landscapes through which he tramped, made his present situation seem all the more hard to bear.

Yet he was grateful for his Australian childhood, youth and young manhood. He considered that it was experience of the outback that made Australians ideally suited for work in the Antarctic. After all, Australia itself had been the site of great exploratory expeditions in generations not long before his own. Between 1840 and the ill-fated expedition of Burke and Wills in 1860, men like John Eyre, Charles Sturt, and John McDouall Stuart led expeditions into the interior of Australia. And though those intrepid travellers dealt with desert, heat, and drought rather than snow, ice and wind, nevertheless Mawson felt he belonged to the same tradition.

These were men who had achieved position, fame, and honour by leaving their respective birthplaces, and forsaking cities to tramp the outback where no white man had been before. Mawson thought wryly of his trips to England before the expedition. The loneliest he had ever felt in the company of men was in London. And although he had enjoyed his brief trip to Yorkshire to visit relatives and to re-acquaint himself with the village of his birth, he had not felt at home among the social distinctions and niceties that were somehow mirrored in the small and intricate divisions of the rural landscape.

Now he was experiencing an entirely different kind of loneliness. He thought of the editorial to the first number of the *Adelie Blizzard*. It was entitled 'Marooned' and referred to *Robinson Crusoe*. Mawson could now envy the hero of that tale in a different way from when he was growing up. As a schoolboy, Mawson, like his classmates, had been taught to look on Crusoe as an example to be followed: an adventurer, a man of ability and resource, of economy and prudence; a man who worked hard and believed in Providence; a man who in the end declared new territory for England. But now it was Crusoe's lush island that Mawson envied, and all the sources of food that the fictional hero had at his disposal.

Like Crusoe, Mawson felt that he could bear loneliness a little longer in order to ensure his survival. But cold and hunger were the real threats to his endurance.

Still there was no light. So Mawson huddled deeper in his bag to wait for the dawn. All he could do was to hope for a good day, and look forward to some more thin broth for breakfast. As he tried to compose himself to sleep, his mind wandered back to Crusoe on his island, believing in the Providence that eventually saved him.

• • •

Robinson Crusoe was an immensely popular schoolboy text in the late nineteenth century. Published in 1719, it inaugurated a tradition of masculine adventure fiction which not only propagated popular imperialism, but also played a massive part in the definition of that masculinity which made the Empire and its continuation possible. A host of writers in the late nineteenth and early twentieth century, following Defoe and Sir Walter Scott, wrote adventure stories for men and boys. Among the English writers whose popularity and influence are beyond question are Captain Marryat, W. H. G. Kingston, R. M. Ballantyne, H. Rider Haggard, G. A. Henty, John Buchan and Robert Louis Stevenson.

Most of these authors were well known in Australia between 1880 and 1913, but there were also Australians who wrote adventure fiction. Rolf Boldrewood is perhaps the most well known of such authors, but there were others, Ernest Favenc, J. F. Hogan, David Hennessy and G. Firth Scott among them, who drew their inspiration from Rider Haggard, particularly from *King Solomon's Mines*.

These books play a part in a wider cultural pattern of the late nineteenth century. From the 1870s onwards, there were fears in England of racial decline and decadence. People looked to the colonies, to the men of the 'frontiers', to provide a model of reinvigorated empire masculinity. In their fantasies of white masculine power and authority in Africa, in India, in the South Seas, in Canada and in Australia, adventure novels provided both a reaction to and remedy for anxieties about imperial decline. In their exaltation of the warrior-explorer-engineer figure and their celebration of action over the primacy of feelings, they provided a powerful model not only for the shaping of boys' beliefs, but also of their identity.

The library of books at Winter Quarters, Cape Denison had an extensive collection of works about polar exploration as well as several other romance and adventure titles including Rolf Boldrewood's *The Miner's Right* and Robert Service's *The Trail of Ninety-Eight*.

There is a close relationship between this fictional material and various aspects of Mawson's life and experience. In *The Miner's Right*, Hereward Pole, the younger son of a decayed family, leaves England to seek his fortune on the goldfields of New South Wales. The ostensible motivation for this endeavour is the hand in marriage of Ruth, the daughter of an English aristocrat. The book dramatises the conflict between different national mythologies: English aristocratic attraction to maintenance of position, property, and stability is opposed to the egalitarian idealism of the goldfield 'diggers' who represent enterprise, social mobility, and fluid wealth. Hereward or Harry Pole, as he styles himself in Australia, is caught between these incommensurate ideologies and identities.

Although Mawson left England at the age of two, and

was only aged eight when *The Miner's Right* was published, he shared in something of these conflicts. He was the younger son of a family that, if not exactly decayed, was certainly in need of some energetic enterprise to save it from such a fate.

• • •

Grit. Bending forward into the wind, Mawson clenched his teeth and hauled upwards. The sledge hardly moved. The gradient was severe and the snow soft. He gulped great lungfuls of the abrasive air and tried to stop the giddiness that assailed him as he willed himself forward. The day's struggle so far had constituted a series of battles up the snow slopes. He had ascended by stages: a climb would be followed by a small levelling before the ascent began afresh. The physical effort was such that it threatened to overcome Mawson's mind; it was difficult to think of anything but the pain, the exhaustion, the breathlessness. He had to fight to resist the easy and early capitulation which his body cried out for by inventing mind games and promising himself rewards.

For every twenty steps forward he gave himself a 'perk', either the smallest bite from a chocolate stick, or a raisin. He still had a few of each left and carried them in a pocket of his burberry jacket. As he paused between each step, gathering his breath and summoning his strength, he repeated the word 'grit' to himself and told himself about 'character'. They were words that his mother had bequeathed to him like precious stones, and the thought of her and of her example resonated with every repetition. He would not disappoint her as his father had done. He would prove to her and to Paquita the strength of his body and mind.

So he toiled on through the monotony of snow and ice, which was broken only by the occasional patches of blue that insinuated themselves through the cloud cover. He felt himself to be a tiny black speck amid the massive grandeur of the icy hills. But the colour in the sky always cheered him, reminding him of the great skies of Australia with their feeling of limitless possibility, their clarity, their essential azure. He would see them again. He would not be cheated. Grit and character. Grit and character. He would get home and make a place for himself; he would fulfil his mother's ambitions for him and restore the family to the eminence it deserved.

At the beginning of the seventh hour of pulling, he heaved forward again; he estimated that so far the day's journey had been two to three miles of zig-zag progress. He would pull for another hour before the promise of rest.

• • •

Douglas's father, Robert Mawson, had inherited a farm in his native Yorkshire, but had been educated in the classics at York Grammar School. A decided predilection for Latin and Greek did not prove a particularly successful grounding for entrepreneurial activity: Robert went bankrupt, losing a large proportion of his inheritance. When Robert's mother died in 1884 (twenty years later than his father), the Mawson farm was sold and the family emigrated to Australia. But hope and effort were no substitute for good sense or expertise, and Robert squandered more money on failed ventures until he was obliged to surrender his independence and went to work as an accountant for a timber merchant.

Robert's two sons, William and Douglas, were brought up in an atmosphere of considerable struggle. Their mother, Margaret, the daughter of a Manx industrialist who was chairman of directors of the Foxdale lead mines, was determined that her sons would learn thrift, a good head for business and, if possible, a profession. Margaret would allow Douglas to buy household goods and especially foodstuffs in large and cheap quantities and then sell them back to her in smaller amounts, presumably keeping the profit for pocket money. She also took in boarders to help out with her husband's ailing finances. Among these people was one Arthur Carpenter, the son of the last whaling captain to operate around New Guinea and Northern Australia. The young Mawson is said to have been thrilled by his stories of adventure.

Economics, science and adventure were inextricably entwined in Mawson's life and career. The academic disciplines that he studied, including engineering and geology, were those that most directly related to mining. From early on Mawson viewed his expeditions into outback Australia no less than the Antarctic as the means to both pure and applied research. The romance of adventure was directly related to the pursuit of wealth *and* the re-establishment of social position. For however much Mawson may have acknowledged the Australian ideal of egalitarianism, he was also aware of his inheritance and the efforts his mother had made to ensure that her children remained members of the upper classes.

By the end of 1910 Mawson had further incentive to consolidate his social position. He asked G. D. D. Delprat for the hand of his daughter, Paquita, in marriage. The Delprats were a wealthy family of Dutch extraction, and on the declaration of their mutual affection Mawson felt it necessary to write a long letter explaining his antecedents and his prospects to the man whom he hoped would consent to be his father-in-law. In the letter Mawson is anxious to establish that he comes from an old landed, if not aristocratic, Yorkshire family. But at the same time he declares a distrust of the 'useless' classical education of the English upper classes which is blamed for his father's business failures. Mawson goes on to speak of his mother's 'grit' in dealing with her lot. Not unnaturally, Mawson wishes to persuade Delprat that he has the useful education and determination which will enable him to keep Paquita in the proper manner.

European nostalgia vies with colonial enterprise, just as an attraction to aristocratic values are in conflict with Mawson's attraction to egalitarian and democratic idealism. On the one hand, Mawson can write in a notebook of 1908 a series of Latin tags including *Omni personam delectes et discrimine remoto* (every distinction between persons being laid aside) and insist upon equality of effort and reward between the members of his expedition—'everybody must be on an equality' he writes—but on the other hand he feels that some people are manifestly superior to others.

Mawson's distinction between those who can do things and those who can't sometimes became infused with prejudices of race and class. Sailing second class to

England in 1909 on the SS *Mongolia*, Mawson confided tersely to his notebook that he 'was not much struck with the passengers as a whole'. Writing to his nephew some years later about taking passage from Marseilles he writes: 'You will find the people on board very rough and very dirty stuff, mainly Jews, Italians, Greek and Maltese.'

Mawson's attraction to Paquita might also be construed as at least in part involving an element of northern European nostalgia, if not class ambition. Born in Spain of Dutch parents, Paquita went to Europe while Mawson was in Antarctica. During his absence she wrote him a series of passionate love letters which also inevitably reveal something of her values. In one such communication Paquita writes of her 'patriotism' towards Holland, remarking that it would be very difficult to return to live in Adelaide if it were not for their marriage. She goes on to express their 'fellow-feeling' and say that when they are married, their house 'will be *the* house in Adelaide, in Australia'. In another letter written on her return from Europe, Paquita speaks of feeling the 'emptiness of Adelaide': 'Adelaide seems a small village and the working class are worse than ever.' From Melbourne, Paquita writes of learning short-hand and typing: 'I started in Adelaide and liked it but the college here was too dirty and common for words so I left . . .'

Mawson idolised this young woman, who embodied something of a European aristocratic tradition in the social 'emptiness' of Australia. Paquita represented not merely romance but also position, property and stability, as opposed to the egalitarian adventures whereby wealth was to be wrested from colonial initiative and enterprise.

In a letter to Paquita from the Antarctic, Mawson writes: 'Do you know that had I not lived in the Twentieth century I might have been something very different. A Crusader or a Buccaneer.' This seems at first sight to be out of character with the immensely practical Mawson. But it does acknowledge something about his conception of his own masculinity. The crusader is the knight riding forth to conquer the heathen in the name of Christ; he is also a commonplace figure of the Victorian re-invention of the Middle Ages, a figure redolent with imperial ambition and significance. The buccaneer is the independent brigand who wins wealth by courage, initiative and daring.

22 January, 1913

MAWSON FELT THE SUFFOCATING weight of icy water pressing down on him, as if his lungs would burst. But through the dim, green, underwater light there he saw the object of his dive: a wooden crate crammed full with bottled jam. He strained and strained to lift the crate from the clutches of the kelp in which it was entangled, but it remained steadfast. The cold, the awkward and abrasive texture of the wood, the hunger that impelled this risk, coalesced into nightmare. Breathless, desperate, almost weeping with frustration, Mawson surfaced through his dream to wake shivering and hungry in another bleak dawn.

Regaining his composure a little in the familiar confines of his canvas shelter, Mawson wondered at the way the mind re-configured experience in dreams. Landing stores at Commonwealth Bay, they had dropped just such a crate and, finding that a part of the stove was missing, he and another expeditioner, Laseron, had set off in the whale boat at low tide to retrieve the lost case, thinking that it contained this piece of equipment. They located the lost crate in six feet of water and, after several failed attempts to raise it by means of hooks and other objects, Mawson had taken his clothes off and dived in to raise it. The temperature of the sea was 30°F—below the freezing point of fresh water—and the air temperature was much lower. It took Mawson three attempts before he managed to haul up the case. For all his trouble, it did not contain the missing piece for the stove but was full of jam.

At the time, Mawson had thought little of the incident, though he was aware it had caused some of the other men to look at him with new eyes. As for himself, he'd run to the hut with ice forming on his skin to get dressed and warmed as quickly as possible.

But the dream turned everything topsy-turvy; he had been frightened and oppressed by the depths, unable to succeed in his quest, and had awoken with a rush of emotion that was difficult to contain. Mawson hated this feeling of anxiety bordering on panic that was becoming so familiar to him. In some ways it was worse than the physical discomfort because it threatened to unman him, to reduce everything to murky subterranean uncertainties which his reason, his science, could not contain or control. That was when terror took hold.

The first time Mawson had experienced anything like this was on the *Aurora* in the pack ice. Having survived a storm that threatened to sink the overloaded ship and finish the expedition before it had begun, the party sailed on to land at Macquarie Island where they erected a wireless mast. This was to be the first expedition to attempt to maintain radio contact with the outside world, and this was to act as a relay station. From Macquarie Island they sailed on south. But then they became hemmed in by pack-ice and fog, and it was at this point that Mawson had begun to feel anxious.

The plan was to land three parties along the coast of Adélie Land so as to explore as much of it as possible. The disposition of the ice made it impossible to land the Eastern party, however, and, as the days went by and coal was progressively diminishing with no landfall made, the whole expedition was in jeopardy. It had been a terrible feeling. After all the time and effort of the preparations, the grubbing for money in London, the responsibilities and the risks, for all this to end in failure was too much. Mawson could not contemplate the humiliation. Was he, after all his early achievements, going to end up like his father—a man of promise who through misjudgment or sheer bad luck, fails and brings hardship and disappointment not only on himself but also on those who love him and depend on him?

He had spent sleepless nights tossing in his bunk, sinking under the weight of care. His engagement to Paquita also inevitably became entangled in the unravelling nets of anxiety. To bring dishonour or ruin upon her was unthinkable. What would her family say? After all, Mr Delprat had urged him to give up the thought of the expedition for the sake of his daughter. To go back with nothing accomplished, to go back to debt and disgrace, these were horrible thoughts.

The extremity of his feelings had alarmed him then as they alarmed him now. He must not give in. Providence had saved the ship, then guided him to Commonwealth Bay. Now he had been spared the fate of Ninnis and Mertz. There must be a purpose. Mawson remembered the terrible trek back from the South Magnetic Pole with David and Mackay. He had kept going then and believed in David's idea of a benign Providence when they had eventually been rescued. Lord Avebury's essay, 'On Religion', came back to him. The intricacies of theology were unimportant. You could tease yourself forever in those depths, or drown in the deeps of the self.

Geology was science—you could read the layers of the earth precisely. But it was the bright clean air and the dazzling sublimity of the snowcape's surface that spoke of God, just as it was what a man did that mattered, not the depths from which that action sprang.

So Mawson talked to himself, holding himself together as best he could, lighting the primus for his miserable breakfast, and bracing himself for the day's march ahead.

• • •

Mawson tugged hard, stumbled and fell to his knees in the soft snow. He remained on all fours, gasping and wheezing. The way was uphill and the surface poor. At 11a.m. the misery of waiting idle in his bag for the weather to clear had given way to the misery of the march. Panting and heaving his way forward, he knew that he could not afford words like 'misery' however insistently they impinged upon his consciousness. The pain in his feet, legs, shoulder and stomach had to be dealt with by stressing the positive in each weary step.

Above him a weak sun lay under high cloud, but looking northward he could see towards the coast violent streaks of cirrus indicative of high wind and bad weather. This spurred him to further effort. He must go on while he could. Each drag forward took an immense effort, with the battle raging in his head between dwelling on the terrible weariness and bodily hurt and the idea that pain did not matter. In order to inspire himself he told himself that his brother, Will, was watching and waiting at the top of the slope. With gritted teeth he forced himself onwards.

Sometimes he was blessed with a few uninterrupted steps when the sledge moved smoothly and his feet found their own way. It was in such intervals that his mind could drift and he found himself wondering what kind of welcome he might expect if he got back to Australia. He worried about his parents. His mother going blind, living alone in her Sydney boarding house, supported only by visits from Will and his wife; his father at last trying to find some stability and financial security through his business in New Guinea. The family needed him to succeed. He had written to Delprat of his family's stubborn determination. He would exemplify this, or die demonstrating it.

Another stumble interrupted the train of thought. He pulled himself along on all fours. Any movement forward would do. Today, if the weather held, he would make five miles. And that would bring further hope of survival. He tantalised himself by remembering the flavour of roast lamb baked with rosemary, roast spuds and peas. Such simple pleasures would never be taken for granted again.

The world beyond his vista of snow and ice, the weak light and scudding cloud, seemed like a fairytale in which Paquita appeared as the princess. He cursed himself for a fanciful fool and forced his aching limbs into the resistant slush. Paquita, Paquita. He repeated the name he found so exotic; the word he savoured as an antidote to misery and despair. Paquita meant harbour, haven, rest.

As he wrenched himself onward, he dismissed the whispering doubter who questioned the substance of his dreams and wanted to know what reality was signified by the magical name. He treasured and protected the sound in his head. Warmth in the cold. Comfort for pain. Ease for all burdens.

He took another rasping breath and dragged upwards.

・・・

Paquita, meanwhile, was composing letters into a vast silence. She received some correspondence in April and May 1912 which Mawson had written on the voyage to Antarctica and during their first days after landfall. Some of this can have been scant comfort to the young twenty-year-old girl whose fiancé had imposed a fifteen-month separation on their relationship. Her father had begged Mawson to reconsider his expedition in the light of their engagement but found his future son-in-law intractable. With little tact Mawson told Paquita just before the *Aurora* left Hobart that he was going away far happier than last time, when she was not awaiting his return. Further letters followed telling of the tribulations of the voyage south and asserting the role of Providence in guiding the ship through storm and pack-ice to safe harbour. On arrival, Mawson could focus more sharply on his beloved. He thinks of her journeying in Europe, and of their future together. But this speculation is not entirely free from insecurity:

I hope Paquita occasionally thinks of Douglas. Perhaps it is your love warmth that already shades me from cold, for I doubt if I feel it so much as the last time . . .

I can almost fancy myself with you in Paris or London. What fun it will be—or will you be bored with me? I sometimes think that I am much better out upon a lonely trail, for nature and I get on very well together. There within gunshot is the greatest glacier tongue yet known to the world. No human eyes have scanned it before ours. What an exultation is ours—the feeling is magical. Young men whom you would scarce expect to be affected, stand half-clad without feeling the cold of the keen blizzard wind and literally dance from sheer exultation—can you not feel it too as I write—the quickening of the pulse, the awakening of the mind, the tension of every fibre—this is joy.

It was probably with some relief that Paquita reached the words, 'I live for you.' She could be forgiven for wishing that the passion, exultation and joy that Mawson expresses about the landscape had been expended upon his thoughts of her.

There is uncertainty on Mawson's part here: will she be bored with him? Does she still want him? The two sides of himself, one following his mother, hankering after the fine life of European civilisation as experienced by the wealthy and powerful; the other, inherited from his father, craving a life of adventure on the frontiers, are in tension. He is living and expressing the conflicts inherent in colonial masculinity.

In the letters she wrote to Mawson during 1912,

Paquita shows no sign of conflict in her feelings, but she worries about her fiancé's. On 1 April, Paquita still hadn't received Mawson's letters and no wireless messages had come through, but she has no doubts about the temperature of her own feelings: 'If you were only here now how your Paquita would warm you ... My love, my love, how I miss you. I close my eyes and lift up my lips but feel nothing. How very far you are.'

On 10 May, having received Mawson's letters, she wrote again, knowing that her letters would have to wait to be sent via the *Aurora* in November. Her letter is friendly and positive; she wishes she'd been there to help during his time of worry before they landed. She acknowledges how different their lives are at present. There is chat about her doings in Holland. But in the midst of this affectionate amiability, one can detect the whispers of doubt that are assailing her. 'Douglas, dear,' she asks, 'aren't you glad we've got each other?' This question is the prelude to a longer outburst at the close of the letter: 'I'm sure you don't love me as I do you. Women always love the most and miss the most. Well I wouldn't like you to miss me as much as I do you. With my whole heart and my lips your Paquita.'

That the intensity of her passion was unreciprocated was not her only worry in this letter. By now there was news from the Antarctic that the Scott expedition was to be away for another year. 'Don't you go and stay away another year!' she writes. 'You're under contract to return next year less than a year from now.' Paquita attempts to remain as cheerful as possible, but the anxieties of her situation are manifest.

As month after month passed with no radio messages,

Paquita could not know that the appalling wind conditions at Commonwealth Bay had defied all gallant attempts to erect the wireless masts. Still the time passed. In mid-October she wrote another letter, knowing now that the time was relatively short before the *Aurora* went south again, but that she still could not expect to see her fiancé until the following April. The passage of time has not diminished her ardour. Paquita begins and ends her letter with protestations of affection, and assures her beloved that he is coming back to 'the warmest and lovingest heart that ever beat for its other half: I can almost feel your arms round me and involuntarily as I write lift my face to yours.'

On 26 December the *Aurora* sailed from Hobart carrying Paquita's letters with it, and after another difficult voyage arrived at Cape Denison off Commonwealth Bay on 13 January. As Mawson toiled up the snow slopes, cold, hungry, and fearfully alone, her words were waiting for him at the hut.

• • •

Mawson crawled around the flapping canvas of his shelter, laboriously weighting its edges with snow and ice blocks. Ice and snowflakes danced dizzily before his eyes. He was faint with fatigue. The process of hacking the ground for snow and ice had taken an hour and a half. He knew he was close to collapse and had to keep stopping to take huge gasps of air. Then the world would stop spinning drunkenly before his eyes and he could continue his crawl.

At last it was done, and Mawson could throw the

canvas over the makeshift tent-poles, unpack his equipment and lie gratefully wheezing, trying to concentrate on the feeling of accomplishment rather than on the utter weariness he felt. It was as if every bone in his body had been strained to breaking point; every nerve was alive with pain. Yet it had been a good day. He had staggered for five miles. The weather had remained fine. He had had a clear view of the sea towards the north north-east. There was hope.

When he had gathered his breath and his wits, Mawson began the process of making some food. Because he felt light-headed he decided to make more hoosh than usual, mixing pemmican, dog meat and biscuit into a grand concoction. When these preparations were complete and he was avidly spooning this luxury into his belly, he felt his spirits begin to revive.

Then sucking a stick of chocolate he began his penultimate task for the day, stowing his rations back into his bright orange ditty bag. Here was another talisman. Paquita had sewn these bags in various vivid hues and Mawson had distributed them among the sledging parties as a homely reminder of the coloured world that awaited their return from this land too often reduced to blank white or, worse still, grey. The men had christened them 'Paquita bags'. Every night as he carefully unpacked and packed the life-giving nutrients, he felt satisfied that they were carried in this bag sewn by her hands.

Invariably thoughts of her hands would turn his mind to that velvet night on the verandah in Brighton. Under the summer night's magical spangle of stars, with the white sand gleaming away beyond them and the soft rhythmic hush of the waves providing their lullaby, he had taken her hand in his and asked, haltingly, if she would consent to be his bride. They had kissed and turned hand in hand to gaze out, enfolded by the breathing mystery of the world, and they had seen a sea bird swerve in its elegant flight, wings tipped with moonlight silver, an ornament to their love.

Mawson sighed as he settled into his sleeping bag. He returned to the practical, to the biting facts of his situation. He had little energy or inclination for his diary entry tonight. He scrawled the account of the day's progress: 'Sledged uphill terrace after terrace. Very soft shifting snow, or else I would have done better.' He noted the wind directions and the compass bearings he had followed and concluded: 'Providence has been very kind today—must measure ration. Snow is so soft and deep.'

He put the journal under his head and as he prepared to sleep amused himself by planning his ideal house. He would design his own, for he was fussy and knew what he liked. Anyway, it was fun to exercise the mind in such practical schemes.

23 January, 1913

MAWSON BREAKFASTED LAVISHLY. Sunlight could be discerned through the flap of his shelter. This good news had determined him to set himself up well for another five-mile day. The food of last night augmented by another half tin of pemmican, a biscuit with plenty of butter and tea made him feel optimistic again. It was as if he could feel the effect of this fuel in terms of renewed hope and energy.

Mawson had always been energetic; he hated sitting still doing nothing. He was impatient. To be up and at life was his temperament. But there were times on this journey when he had felt so starved and beaten that this inclination had almost died within him. But today the weather seemed to be a signal that Providence was going to see him home. He had done his sums with the food and it still seemed possible to survive if the weather remained in his favour.

Sometimes he imagined his reception at the hut. He felt fairly confident that most of his fellow expeditioners reacted positively towards him. There was the obvious exception of Whetter and there had been a few niggles with some of the others, but surely there would be the rejoicing and fellow feeling that comes when men have shared hardships together. The men would be grieved over the loss of Ninnis and Mertz, but he could not be blamed. He would explain everything to them, and all would be well.

Thinking of the hut, however, brought with it a further worry. If he did get through, would the ship have waited for him? And, if it had, what would be his reception at home? The expedition had done its best, and there would be scientific results to show. His story might be worthy of some interest. But there were aspects of the expedition that had not been as successful as he had hoped. The extremity of the conditions at Commonwealth Bay had meant that next to no oceanographical work had been possible and the wireless had remained inoperative. The hope of finding accessible mineral wealth was dashed by the terrain and ceaseless gales. And he had not got as far as he would have liked on the sledge journey. He wondered how the others had fared. He hoped they had not met with the same completely inhospitable country that he'd been forced to traverse. Perhaps there would be some significant gains in mapping.

However he looked at it, the expedition was certainly not going to make him rich. Mawson thought of the letter he had written to his prospective father-in-law and the poem he had quoted in conclusion. He wanted to assure Delprat of his feelings for Paquita, but also of his recognition that she deserved to have a husband who could provide for her properly and keep her station in society. He needed to justify the expedition. The words came back to him as he packed up his gear and prepared for the day's labour:

I'll say: 'Here's bushels of gold, love,' and I'll kiss
   my girl on the lips;
It's yours to have and to hold, love.' It's the proud,
   proud boy I'll be,
When I go back to the old love that's waited so long
   for me.

Well, there would be no gold, and perhaps only the wreck of a man. But when they heard his story, surely they would be merciful. Paquita had thrilled to his stories of the Shackleton expedition. If he survived this ordeal, he hoped she would see the romantic side of it; he was, after all, *living* the adventures of Robert Service's verse. Whether the men and governments who had supplied the financial backing for his undertaking would take a similar view remained to be seen.

But these were minor considerations. His task for today was brutally uncomplicated. To heave, haul, crawl five more miles towards Aladdin's Cave and safety. That was the imperative, the challenge. Mawson braced himself to begin again.

• • •

Although he saw himself as an idealistic scientist, Mawson was not the kind of man who wished to spend his life day after day in a laboratory in the company of microscopes and test tubes. On the contrary. The blunt scientist saw part of himself in terms of the adventurer of popular imagination. His shyness and reserve kept this hidden from most of his fellow expeditioners, but it surfaces in his strong attachment to the masculine romance of Robert Service. This is the literature of action and sentiment. The sentimentality is the obverse of the hard, tough persona projected through the verse. It is also a sentimentality at the service of an economic idea. Service's writing, however light in tone, weaves together ideas of heroic masculinity, colonial derring-do, economic advantage and love: love is gained by 'real' men who have proved themselves, made a fortune, and therefore are worthy of their beloved. Science, for Mawson, was inextricably interwoven with such ideas.

The theme of economic advantage was always present while Mawson was planning his expedition. Following his failed negotiations with Scott and Shackleton, when Mawson began to seek financial support for his own endeavour in earnest, he was obliged to seek help from both individuals and governments. In the various letters and documents he wrote outlining the aims and advantages of the expedition, it is clear that there is a very close relationship between science and adventure, if adventure is understood to be about venture for economic gain.

In a 'Brief Summary of the Objects' of the expedition, Mawson lists the six scientific fields of endeavour involved and the value of these to Australia. He points out that mapping and surveying the Australian quadrant of the Antarctic continent must be of economic value in the near future; that geological work has its commercial interest: coal is known to occur and minerals are to be expected. Mawson linked the biological and oceanographical work of the expedition to potential economic gains from whaling, sealing and fisheries. The magnetic charting of the region, he argued, would be of direct benefit to Australian shipping. Only the meteorological work is left without an economic rationale.

Mawson also knew how to appeal to different audiences through national or Empire sentiment. In proposals aimed at English audiences, Mawson shrewdly downplayed the strictly Australasian aspect of his plans and emphasised imperial ambitions. He speaks of wishing 'to raise the Union Jack and take possession of this land for the British Empire'. The rivalry of other nations has to be opposed. He says that the solving of the South Magnetic Polar problem will remain a permanent achievement to the British nation.

But when addressing an Australian audience in search of support, Mawson, though occasionally playing on both nation and Empire, emphasises the nationalistic aspect of his enterprise. He writes that 'it is intended to raise the Union Jack and the Commonwealth flag' on the new Antarctic territories. He suggests that the geographical position of the part of Antarctica he wants to explore is 'a birth-right of Australians'. The publishing of scientific results from this region 'is earnestly sought after by Australian scientists. We do not want the intervention of foreign expeditions.' The South Magnetic Pole needs to be 'thoroughly ear-marked for Australia', but before this can be achieved more work needs to be done in the area so that 'the accomplishment of first reaching the South Magnetic Pole' may be regarded as a 'national achievement'.

Mawson wrote these appeals in a period of intense rivalry between the major powers which had resulted in fears of foreign invasion. In England, fiction and non-fiction dwelt on the possibilities of invasion by Germany.

In Australia from 1905 onwards there was a similar spread of writing about the threat said to be posed by the Japanese. Although Mawson is not explicit about this in any of his appeals, the idea of a foreign power claiming land that was closer to Australia than Melbourne was to Perth was not palatable. That a Japanese expedition was also in preparation at the time cannot have dampened such fears. Certainly the newspapers were very aware of this angle. A cartoon appeared at the time which showed 'Science', dressed in Grecian costume, pointing at a figure in Antarctic costume while delivering the following homily to a comfortable Australian with pipe in mouth: 'You are prosperous, the seasons are good, so be a sportsman and back this young man against foreigners.' The 'foreigners' were the Japanese. 'Science' could evidently be made to lend high-sounding support to some fairly crude and less altruistic motives of political and economic gain.

• • •

Mawson cursed as the sledge harness dragged him sideways, nearly knocking him over. The early sunshine had disappeared, blotted out by low cloud and swirling drift. The wind was so strong that it kept capsizing the sledge; each time it happened Mawson was wrenched around and had to go through the arduous process of setting all to rights again. His sense of disappointment in the day was strong, but in desperation he persevered.

The seething cauldron of snow through which he staggered blotted out the few features of the snow plain, and navigation by means of the sastrugi was rendered impossible by the amount of snow obscuring their trends. Mawson attempted to orientate himself by means of the wind direction. But that seemed to veer and shift alarmingly, making him uncertain of his progress. His goggles clogged with snow. He stopped again and wiped them, then proceeded, goaded on by the thought of all the food he'd eaten; he must translate that energy into further miles towards safety.

Over the last few days Mawson had found himself thinking aloud. Now he talked to himself about the impossibility of the weather and the need for fortitude. 'For Paquita, for Australia,' he chanted as he put one foot in front of the other. Occasionally thoughts of his earlier ordeal trekking to the South Magnetic Pole on the Shackleton expedition came to his mind. At least this time it was all up to him. Then he'd had to fight and negotiate with his companions to keep them going. Now it was Douglas Mawson, D.I., Dux Ipsi, the leader himself, who had to show what he was made of. Nobody would have cause to say that he gave up, that Australians couldn't do as well as anybody in Antarctica. He recalled Scott and the priggish Dr Wilson. Scottie was all right, but Mawson had not thought much of Wilson. He looked ascetic and constipated, as if practising for sainthood. Mawson grinned to himself. He couldn't imagine Wilson in one of Service's poems. But he could and did imagine himself. He was made of the same tough stuff.

So he told himself the story of his own strength, endurance and determination, as he struggled forward through the blizzard. The snow was soft and slushy, impeding the sledge, but proving comfortable for his wounded feet. Here then, in the middle of the storm, was a blessing. He was no longer trying to drag uphill and he had left the nightmare of the glaciers far behind. There was plenty to feel positive about. So drag on, drag on. He told himself it didn't matter if every step hurt, that physical pain was meaningless. The only thing that mattered was to keep going.

He allowed visions of his ultimate goal to come to him. He thought of the feeling of triumph he would have as he sailed into Port Adelaide on board the *Aurora*. He imagined the scorching heat of an Adelaide summer, saw Paquita, a bead of sweat just discernible above her lip as she tilted her head towards him to be kissed. And wouldn't there be a fuss? There had been last time he arrived back from Antarctica. The students had met him at the station and pinched a barrow in which he was chaired triumphantly along North Terrace to the university. A policeman had chased the boisterous procession in order to repossess the vehicle. Despite Mawson's self-conscious shyness in crowds, he'd felt the exhilaration of success and recognition.

He would feel them again. 'For Paquita, For Australia.' Step by step, zig-zagging home.

24 January, 1913

AS SOON AS MAWSON SETTLED his weight on the sledge and lifted his feet from the ground it began to move downhill through the swirling drift. The wind and surface were for once favourable to him, though the direction of this eccentric glissade was less certain. Still there was an exhilaration in the movement and the thought of ground gained for such little effort. There was also a strange excitement in the feeling of being slightly out of control. He had no steerage, the sledge making its own way propelled by the mighty force at his back. When a particularly strong gust accelerated the descent, Mawson let out a whoop of glee. He reminded himself as he did so of the youngsters he had seen tobogganing when he went to Yorkshire in the winter of 1910.

He had been to visit his relatives and had watched boys and girls at their happy sport on the snow-covered slopes around Rigton, under the brooding outcrop of Almscliffe Crag. Now he was a child again in his desperate enjoyment, and he thought of that other Yorkshireman, the ebullient Shackleton, who could also be so childlike in ways. How he would have enjoyed this headlong career through the snow. Scott, on the other hand, was a different proposition. Mawson could not imagine him unbending enough to act childishly, even if he was alone.

Mawson was propelled from such thoughts as the sledge hit a ridge and turned onto its side, catapulting him into the soft snow. Struggling to his feet, gasping, wiping snow from his face and goggles, he had to haul on the harness to restrain the sledge.

The wind was now so strong that Mawson worried about the practicalities of erecting the tent. The gale was increasing in force and it was taking him all his strength to hold the sledge. This did not augur well for the struggle with poles, madly flapping canvas, and ice blocks that needed to be dealt with before he could think of a minimum of food and heat. He reckoned that he had moved five miles already helped by the wind, and so decided to camp there and then. He weighted the sledge with loose snow before releasing himself from the harness, and then fortified himself with a stick of chocolate. It was going to take all his strength over the next few hours to arrange his camp.

・・・

Mawson lay in his bag at last, the day's struggle over. He had spent the last hour of activity treating the palms of his hands with lanolin, for most of the skin had now come off them, making all his activities that much more painful and arduous.

Putting up the tent had taken two and a half hours and by the time it was up the tent had filled with snow because of the force of the wind, and all his gear had been buried. On hands and knees he had had to excavate and shovel snow, until he could eventually crawl into shelter from the blizzard.

It was with immense relief that he now lay down to sleep, but rest did not come easily. The noise of the wind, the hissing of drift against the flap and bang of the canvas, produced a cacophony; it did not augur well for the march tomorrow. And he was haunted by his own loneliness.

He remembered all the nights he had spent sleeplessly anxious when preparing for the expedition. He remembered the sense of loneliness he'd had in London even though he was surrounded by people. Human beings he considered were not, on the whole, either pleasant or reliable creatures. Take a man like Shackleton. He had so many admirable qualities—he was tough, brave, enthusiastic, with something of the poet about him—but he was also duplicitous and conniving. You could not trust him with money. You could not trust him at all.

And yet through the knowledge of our own frailty as well as our self-regard we are bound to crave the company and love of our fellows. Mawson knew he was difficult to please. He was fiercely impatient with the inefficiencies, laziness and foibles of other people. Lack of seriousness and enthusiasm irked him, while the inconsistencies of human beings made him long to retreat into the beautiful orders of science.

He thought of Kathleen Scott, the living antithesis of scientific detachment. He had had lunch with her and Paquita in Adelaide after Scott had left for Antarctica. It had been a vivacious affair. Kathleen took command of the situation and the conversation, and Mawson had relaxed so much that he had said more than he ought to of the horrors of his march with David and Mackay to the South Magnetic Pole. He had been flattered by her attention. Undoubtedly Kathleen had focused upon him rather than Paquita, whom she had treated in a benevolent and condescending manner rather as if she were a child. But to Mawson she had listened avidly, and he had glowed at being able to recount his heroic exploits to both women.

Paquita had not been completely enchanted with Kathleen, but this was to be expected, thought Mawson with a grin. Although he was glad he wasn't married to her, he liked her more than he did her husband. Kathleen had some spirit and some brains. Despite Scott's charm, there was something buttoned-up about the man, there was too much of the naval captain used to ordering people about. Shackleton was much more democratic, even if his morality was questionable. But then Shackleton was of Irish descent; Scott, Mawson thought, was quintessentially English.

Mawson's mind returned to Australia. Australians were more energetic, more practical, more forthright than the English; they were at the forefront of the race. His survival would help to demonstrate this. With this thought in mind, he translated the sounds of the storm outside into the comforting reminiscence of the sounds of sea and rain at Brighton, his place of love and dreams.

25 January, 1913

AS MAWSON CAME TO CONSCIOUSNESS the first thing he registered was darkness. This together with the hiss of the drift and the howl of the wind, made his spirits slump. He knew immediately that he would not be able to go on if the weather did not improve.

He forced himself up and out of his bag to take a look outside. The scene confirmed his worst fears. Visibility was about five yards. It was like standing in the middle of a boiling cauldron, the drift seething in massive agitation, blown into anarchic frenzy. Sky and horizons were obliterated. His sledge was half-buried. He could only retreat to his bag.

At first he busied himself preparing food and attending to the boils that covered his body. He dressed his hands with lanolin. His nails were beginning to fall out. Every morning there were great tufts of his hair and beard decorating his bed. He had to resist the rising feelings of panic that came with thoughts of his decaying body. He must discipline himself to think only of what is, not what might be.

As he lay back in his bag, all conceivable chores having been completed, he thought of his previous long journey through snow and ice and sought comfort from the recollection. He had been through hell before in 1908, man-hauling two sledges by relays. They had crawled up the coastline, looking for a way inland towards the South Magnetic Pole. There had been rows, niggles and petty irritations as he and Mackay stumbled on, their progress impeded by the fifty-year-old professor, T. W. Edgeworth David.

David, or 'Tweddy' as he was known, had been Mawson's mentor at Sydney University. The garrulous enthusiasm of the man for all things scientific had spoken directly to Mawson's austere silences, opening up new worlds of possibility in the study of rock and stone. He was Mawson's chosen father figure in the pursuit of a university career and in the absence of his own father overseas.

But the march had changed all this forever. Although in its aftermath they salvaged their relationship, the journey exposed the fissures between the two men so radically that an entire reconciliation was impossible. David's inability to come straight to the point, his indecisiveness, his annoying personal habits, his disregard for the comfort of his companions, his very age which prevented him from sharing the physical burdens, had all caused tension and rows.

Then there was Mackay. He was tough and resolute, but a follower, not a leader. Towards the end, when they were staggering home, Mackay had physically kicked the professor along in the traces. They had been a desperate bunch by then. They had travelled twelve hundred miles and were suffering from malnutrition and frostbite. The professor was occasionally demented, his feet gangrenous. He had staggered on, attempting to walk on his ankles. His suffering must have been terrible. But Mackay was beyond pity and encouraged the ailing man's progress by kicking him along from behind.

Mackay's behaviour was erratic and furious. It culminated in an argument about who should have control of the party's pistol. Mackay told David he was a bloody fool and let out a string of obscenities, saying that Mawson should formally be made leader and that David was insane and no longer responsible enough to take charge of the firearms. There had been an unseemly struggle in the snow before Mawson came between them. Mawson did not want to force David into the ignominy of formally handing over the pistol and the leadership. He was spared this dilemma when they were rescued by the *Nimrod* come in search of them.

Now he lay in his bag alone. The awful memories of their squabbling and fighting all the way home made Mawson feel that things could be worse. That he had survived such privation before gave him hope and strength. His situation now was more extreme. He was physically in much worse case, and he had much less to eat. Still he felt he could endure, if only the weather would help him.

But the wind howled on hour after hour, and the weight of snow began to make the tent collapse around him. He tried to rest so that he would be ready to move as soon as the weather improved, and to this end he exercised his mind by variations of counting sheep. He tried to remember the imaginary menu that Tweddy had invented when they were battling home, their gums bleeding with every mouthful of biscuit and seal blubber. David had promised them such feasts when they returned to Sydney. He called one of them a 'Yorkshire Empire Dinner' in honour of Mawson's birthplace, and another was a special Scotch dinner for Mac.

It took a long time and considerable concentration for Mawson to reconstruct the 'Yorkshire Empire Dinner' in his mind. But in the end he believed he had the full twelve courses down pat. There were scalloped oysters, tomato and turtle soup, boiled English salmon, jugged hare with blackcurrant jelly and champignons, young duck with petit pois, new potatoes and American sweetcorn. All this was a preparation for the main traditional dish of roast undercut with Yorkshire pudding and horseradish cream. Mawson considered the professor's choice of vegetables eccentric, because with the beef and Yorkshire pudding he had suggested Boston baked beans, baked tomatoes and roast spuds. Still it would be good to try this one day.

Then there were sweets. Trifle, omelet au Rhum, cheese soufflé, apple and mince pies with cream, coffee and chocolates. Mawson spread all this before his mind and wondered if anyone who had not been desperately hungry could really appreciate such a feast. It was an imaginative table that would only give joy to those who had suffered the deprivations of near-starvation.

Mawson sucked on a stick of chocolate and rehearsed the menu again. If he could keep going for another week, surely he would reach the hut. Then there would be food, shelter, a bed, the voices of his companions and letters from Paquita, for the ship should have arrived by now.

• • •

Mawson's trek in the company of David and Mackay to the vicinity of the South Magnetic Pole was extraordinarily arduous. They travelled about 1260 miles. Of this 740 miles was relay work, and they dragged huge weights with them. When they set off they were pulling over half a ton, and for the whole journey they never dragged less than 450 pounds. David, Mawson, and Mackay man-hauled for a longer period of time than the members of the Scott expedition, and were in the field almost as long. The total time of their journey was 127 days. By the time Scott's party reached their final camp they had been travelling for 139 days, but only 99 of these had been man-hauling. The one advantage David's party to the Magnetic Pole had over Scott's expedition to the Geographical Pole was that their way took them at first along the coast where they could cache some of their supplies and live off seals, and birds. But for this they would have been unable to survive.

Partly to educate his men and partly to impress them, Mawson had entertained his companions during the winter with tales of his march with David and Mackay and how from the start they were impeded by dreadful terrain and conditions. He told of how it had taken a fortnight of agonising labour to cross the twenty miles of the Drygalski glacier with its chasms, steep ridges and crevasses. But this was not the end of their tribulation. They then had to find a passage up to the high plateau. Several routes were tried before they finally made a successful ascent. Once there, it was difficult to breathe as biting winds scoured their gasping throats and lungs. Their lips, fingers and feet cracked as they heaved onwards, exhausted from insufficient rations.

Mawson told of the blow to their morale when they realised that the Magnetic Pole had shifted further inland than the position previously assigned to it. And the awful necessity of doing sixteen to twenty-mile forced marches in order to reach the coast and any prospect of being saved by the *Nimrod*. The final stages of their journey down the snout of the Bellingshausen glacier were made unspeakably tense by the uncertainty of meeting relief from the ship.

Mawson also told his companions about the difficulties between the three men which he had turned into amusing stories about old Prof. David's follies. How he had come into the tent late and then spent hours arranging and rearranging himself and his socks while alternately sitting on the legs and faces of his comrades. How his pockets had been full of food scraps, books, bonza pocket knife, and how this cramped them all in the night. And then there were his garrulous monologues, quips and quotations to put up with while Mawson and Mackay struggled with all the heavy work.

Mawson had made his companions laugh with such stories. But then in the silence of his own bunk felt guilty for speaking so slightingly of a man to whom he owed such a lot. In the end, Mawson had forgiven David, because he had shown extraordinary pluck in continuing at all. But there was, Mawson reflected, only one leader, and that was himself.

. . .

Mawson lay with his nose only an inch from the encroaching canvas of his tent. The enclosed space made him feel that his sleeping bag was like a shroud, the tent his coffin. He pushed this thought away and made an effort to think positively. He had been in tight spots before and had prevailed. And on those occasions there had always been companions to worry about. Now there was only himself. He worked to make something positive of this. He thought of the arrival of the *Nimrod* off the Drygalski ice tongue, how they had scrambled from the tent whooping for joy, and how David had been trampled underfoot by Mackay in his eagerness to leave the tent.

Taking up his pencil to record the day, Mawson mustered his faith to stand fast against the gale:

A violent blizzard, could not travel . . . Snowed practically all day. Sun gleamed in morning about 9am, but could do nothing by myself in such weather as would blow packages away; besides little hope of striking camp again. I cannot sleep and keep thinking of all manner of things—how to improve the cooker etc—to while away the time. The end is always food, how to save oil, and as experiment I am going to make dog pem and put the cocoa in it. Freezing feet as too little food, new skin and no action; have to wear burberries in bag. The tent is closing in by weight of snow and is about coffin sized now. It makes me shudder and think of the latter for a moment only. I am full of hope and reliance in the great Providence, which has pulled me through so far.

26 January, 1913

MAWSON SPLUTTERED AWAKE, writhing as he struggled against the weighted canvas. He had been dreaming of being buried alive, and as he came to his senses realised that the nightmare had been provoked by the touch of the tent brushing his face. It had collapsed even further than it had last night under its burden of snow. And outside he could still hear the shriek of the wind.

But he was alive. He would not dwell upon coffins. He determined to force his way out of the bag and begin the arduous business of shovelling his way out. He worked purposefully, taking frequent rests to recover his breath and relieve the stinging from his damaged hands. As he worked he looked forward. Once he had dug himself out, he would set up his shelter again and have some food. Then he would decide whether to travel or not.

In his heart he felt that he must try to move on. Last night he had calculated that he only had about four or five pounds of food left, and he thought that he was still about fifty miles from Aladdin's Cave. The equation was still just manageable as long as he kept moving. But he could not afford to eat and stay still. That was the way to ensure that his bag would be his winding sheet, the tent his frozen casket.

This is doing me good. It's warming me up, slowly. But my feet are still cold. I'll try cocoa in the dog pem like I wrote last night. Not a bad idea. More nutrient for less fuel expended. Taste might be strange. But not tasting much anyway. Mucous membranes and saliva glands so badly affected. Yet I imagine taste. Strange that. How much of what we experience is imagination? Can't afford too much in some directions. The tent as coffin not much help. Have to be practical. Can't imagine death. Imagine life. Food. Survival.

He realised he was talking to himself out loud again. First stage of madness? He didn't really think so. It seemed to help him go on. He thought of the professor raving and demented towards the end of their southern trek. Well, he would not be like that. He was glad he wasn't hampered by anybody now, though. Another as strong as himself, Ninnis perhaps, might have helped. They could have used a sail more in the stronger winds because two could put up the tent. But then it would have been another mouth to feed.

Hadn't he always relied on himself? No good getting too close to other people. That's when the rows started. No good getting buried by emotions. No use dwelling on this subterranean stuff. Keep everything clean and clear, like Antarctic weather at its best. That beautiful cool. Yet he felt the need to talk, but you could talk and laugh without giving too much away. He wondered how much he'd given to Paquita.

He dismissed this thought as unhelpful and continued

to burrow his way out of the tent. When at last he crawled free, it was to confront not the idealised beauty of Antarctica, but its terrible destructive power. The wind was still blowing very strongly, maybe up to sixty miles an hour, and the air was full of large swirling snow flakes that blotted all horizons. Still it was good to stand upright for a moment before reverting to all fours in order to continue the work of getting the tent into a state so that he could cook inside it. Then the sledge would have to be dug out.

As he continued with these tasks, Mawson debated anxiously with himself about whether to travel or not. It was a terrible risk. For he had not ever tried to put up the tent in such a wind as this and, if he were to lose the canvas, he was surely doomed. On the other hand, not to move was to chance being buried so deeply that he wouldn't have the strength to get out. There was also the critical question of food and fuel to consider.

On balance, he thought the risk of moving was more attractive than another day of idle waiting, pinned down in the tent. To be moving, to be doing something, was such a relief to the mind, however difficult it was physically. He would eat first, then press on.

• • •

The wind blew fiercely at his back and moved the sledge along with no need of sail. Mawson was obliged either to fling himself onto the sledge and risk coming to a halt, or, in a staggering run, attempt to keep the bow of the sledge from careering into the back of his legs and knocking him to the ground. He also had to try to keep on course. Visibility was virtually nil. The plateau ahead was a seething whirl of snow. He steered by calculating the wind direction in relation to the sastrugi.

The snow flakes were large and hard, hissing and rattling as they struck the sledge and beat against Mawson's leathery countenance. Frequently he felt faint and dizzy. It was a miserable proceeding, made bearable only by the thought that he would cover a good distance during the day if he kept going. There were frequent spills and upsets, after which Mawson would sit in the snow gasping for breath before hauling himself to his feet to go on.

And all the time it was a matter of will and desire driving him on. To die in the snow and ice, defeated by conditions and circumstances, did not accord with his sense of self and destiny. He would not give in. Death, and such a lonely death, was to him so abhorrent that it forced him onward. He felt there was still so much for him to do. He was too young for capitulation. He had not experienced what he thought of as the mysteries of physical love. He wanted children. He wanted to be professor at the university. He wanted a large house in Brighton. He wanted to see his expedition recognised for its accomplishment, not least in showing that Australians could endure where others would have refused or crumpled.

He stumbled, crawled and occasionally sailed forward. Sometimes he was so light-headed that it felt as if he were floating over the ground. It was then that he imagined the gale as the gift of God, the Providential will carrying him forward.

In his thoughts, sometimes close to delirium, the word 'will' repeated to himself made a connection to his elder brother who was now a doctor in Campbelltown. A master at Fort Street School had once said that he, Douglas, would never be the equal of his brother. He was the *second* son, but he would never consent to be second in anything. Will had distinguished himself at school and topped his year in the finals of the medical exam at university. Mawson's academic achievements were never quite so sparkling. He recalled with chagrin his second-class honours in physics, chemistry and engineering. His only first-class result had been in geology. But this had merely made him more determined to prove his own outstanding qualities. He would not be put in the shade by his elder brother; he would not be put into the shade of death and failure.

As the light began to fade, Mawson knew that he faced another massive task before he could rest. The tent had to be put up in the blast. It would be a tremendous fight in which the tent-poles often collapsed, the canvas tore itself away from the ice block moorings, the sledge and his gear were continually buried in snow. Once the tent was up, he had to excavate his sledge and the cooking gear. By the time he had done this, the tent and his sleeping bag were full of snow. He had to start excavating again.

It was midnight before he was in his damp bag, eating his mite of comfort, and wondering if he could possibly go on tomorrow. His wet clothes and accoutrements would not help his starving body, though this night, as every night, he did what he could to treat the boils and abrasions on his skin, and those raw parts where skin would no longer grow. His last task was to scribble in his diary a plain statement of the day's story:

Continuance of blizzard, heavy pelting pellets of radial snow. Wind veered during day from more southerly in morning to SE apparently about noon and not so strong. This wind has been quite a 60mph wind—more I should say.

I got off after noon in dense falling snow & drift and went with the wind. This was a great experiment as I had no idea whether I could put up tent in it. However all went well except that I got into an awful mess—everything saturated with snow, and perspiration, in gear also. It was midnight before I was ready to cook dinner—took a time to put up tent and get snow out of things . . . Surface almost flat, only slight undulations detected—if anything more downhill than up. Surface soft with deep snow but very smooth.

27 January, 1913

MAWSON LAY IN HIS sleeping bag listening to the moan of the wind, which had formed such a mournful reprise throughout the expedition. He wore all his burberries in an attempt to keep warm and dry out. His gear was still wet from the exertion of the previous day and night.

Earlier, Mawson had staggered outside to be met by the same conditions as yesterday—howling wind and drift that had sent him back to the scant comfort of his shelter. He did not feel strong enough to go on. He was not prepared to risk another struggle at the end of which he might not have the strength to put up the tent. He would rest today and hope for better things. A few days' decent travelling and he would reach Aladdin's Cave. But the food supply was becoming critical. He had to eat enough to have sufficient strength to move on. Yet if he gained no mileage he could hardly afford to eat. Compromise was the strategy. A few raisins every now and then. The occasional nibble on a stick of chocolate. A prolonged gnawing at some of the cooked dog meat that was now like hard tack and could give the jaws ample exercise.

Mawson lay drifting in a twilight world between full consciousness and sleep, where thoughts and dreams and memories paraded curious fragments and formed a bizarre collage. The discomfort of his present prompted painful images from the past as if to remind him that he had survived before. One minute he was stumbling through the drift with David and Mackay, the next he was shivering in the heat of the New Hebrides as he sat opposite the native boy, beside the missionary in their shallow canoe, his leg alive with pain. They moved agonisingly slowly down a nightmare river, the jungle encroaching on both sides. At any moment he expected to be attacked by the natives of the place who were by reputation head-hunters. The smell of rotting vegetation, the sight of his injured limb, blackening to the groin, the delirious movement between oppressive heat and the cool of the nights conspired to invoke a feeling of terror and panic in Mawson.

He had had to use all his mental powers to push back the emotional darkness which threatened him, just as now in this tent he had to fight to rehearse the facts of his situation, not to give way to the fanciful and terrible. Investigating the geology of the New Hebrides for the British Commissioner in 1901, he had been exploring a remote part of one of the islands when the accident befell him. He had been chipping at a rock specimen by the side of the river when a splinter hit him in the knee. At first he had not taken much notice, but then the knee began to swell and blacken. A sliver of rock had embedded itself under his kneecap.

For thirty-six hours the three of them rowed back to the expedition's ship, HMS *Archer*, where medical help was available. At the time it had seemed an endless trial. Now he knew better. The process of pain and suffering was never endless, but the mental strength needed to combat these experiences was. Mawson clung to the facts. He was about forty-two miles from Aladdin's Cave. He had a few days' rations left. If the weather moderated he would get there and all would be well. It was no use thinking of the alternatives. He kept them at bay, just as in that boat he had hacked the jungle back from invading his mind.

He thought about the contrasting landscapes and how both were threatening in different ways. The jungles of New Hebrides were like an over-ripe fruit, soft and wet with the smell of corruption. Yet they held an uncanny fascination, exacerbated by the thought of the cannibal tribes with their bare-breasted women. His mind shifted from proud, immodest black sensuality to the white expanses of Antarctica. Here virginal purity of unspeakable grandeur reigned, both enticing and terrifying with its limitless cold. Here there was no one. Only thoughts of Paquita, frightening and enticing. Would she love him in the way that he desired? Would she find him boring and clumsy? Would she threaten to overwhelm him with the ardour of her embraces? Men, he considered, had to conquer the sensual jungle and the ascetic ice through their strength of will and reason. Just as men had to rise to the challenge of women, or be lost forever to the fever or frost of their embrace. It occurred to Mawson that the beaches of those islands, Efate and Santo Spiritu, were as white under the glare of the sun as the snow plains upon which he was now marooned.

Perhaps his leaving on this expedition, his refusing to

comply with Delprat's wishes, was in some ways a test for both of them. If their love, their mutual regard, their affection survived this trial, then surely they could marry with a certain confidence. He needed to believe this. He had used her name as a talisman to get him this far. But lying in the cold trying to fight back despair, huge doubts threatened to betray his trust, as if his memories of that exotic landscape, or his knowledge of the scene outside his shelter, somehow eroded his confidence in love.

He concentrated again on conjuring her and realised that strangely, he could not see her in his mind whole. It was as if he was looking at pieces of a jigsaw that he couldn't quite put together. He could remember the shape of her mouth and teeth, the profile of her nose, the whiteness of her neck; but he could not make them coalesce into a single image, just as he could not make her body whole, pressed against his own on the balcony of his imagination. That she could disappear like this was terrifying. Mawson pushed his thoughts in other directions but realised dimly that she was as exotic and distant to him as those palm-fringed, volcanic islands where he had survived his adventures and returned triumphant, with memories that both attracted and repelled him.

Mawson's mind went home to the university. There he had been happy. Both as a student in Sydney and as a lecturer in Adelaide. They were places of refuge, masculine yet nurturing. There he found an enthusiasm and idealism to look up to, carrying forth the great traditions of the British universities in the Southern hemisphere, learning to add to the scientific authorities of the past. It was a relief to think of such things. To turn from the indefiniteness, the vagueness of emotions and women, to think of the hard certainties of science. It made him feel strong again, strong as he felt in the company of men.

Going from Fort Street School to the university had been wonderful. He entered a world in which he felt at ease, and where his imagination was fired by all he learned of chemistry, physics, engineering, mining and geology. They were the subjects around which he could weave his daydreams of adventure, wealth and discovery. They were the subjects that one could grasp as you could grasp the handle of a shovel to dig. They had the magic of authority passed down from man to man. They were the subjects that would confer power in the world as well as knowledge. In this, Mawson considered, they were the antithesis of his father's classical education which, when confronted with the realities of the world, had resulted in failure and weakness. Mawson knew from the time he entered the university that he was not going to make the same mistakes.

He thought of the camaraderie of students in the lab and on the sports field. Even if you had no special friends, there was always a group of like-minded young men to talk to in a light-hearted way about rowing or the cricket. Even to josh with about girls, just as they had done through the long winter in the hut. He remembered the night they had eaten the cake that Paquita had made, and the men had chorused, 'Who makes top-hole cake?' How shyly triumphant he had felt. He had said nothing, but simply smiled. There was nothing to say. He could not talk of Paquita to the men, but he could bask in their knowledge that he had a fiancée, that he was a sexual success, without ever once having to confront the reality of his feelings or his relationship with her.

This was the comfort of being with men. They were like you. They understood matters that needed no explanation or justification. When Laseron spoke of his first affair, explaining that 'you have an itch somewhere but can't scratch it', you knew exactly what he meant; and when the same fellow dressed as a lubra to Hurley's nigger you could laugh unrestrainedly with the rest of them. It was easy, if less exciting, to be with men.

Mawson decided it was time for some food. It was good to have at least a little activity during so long and weary a day. He battled to sit upright against the weight of the tent, and found his supplies with his diary. He would not have much to say today. Still the discipline of writing had to be maintained, like the discipline of keeping his mind looking forward and positive. There was no room in the writing for self-indulgence. Mawson thought that there was little of the effete artist about him. All those dealings with muddy emotions were a betrayal of masculinity. When he had read to the men in the hut it had always been something improving by manly writers like Service, Kipling and Stevenson.

As he prepared to write his diary he was upset to find again that large patches of his hair and beard were falling out. He wondered how much he was going to lose and if Paquita would still find him attractive if he went bald permanently. Another useless speculation. Better to stay with the facts:

> Same wind, less snow falling, conditions appear to be moderating somewhat at noon. My clothes and bag and all gear wet with yesterday's business, so may not move out. Aladdin's cave should be on a

course between N45° & N50°W, distant about 42½ miles. For the last two days my hair has been falling out in handfuls and rivals the reindeer hair from the moulting bag for nuisance in all food preparations. My beard on one side has come out in patches.

· · ·

Doubtless with the rest of the men, Mawson found amusement, relief and solidarity in entertainments which involved cross-cultural cross-dressing. He had seen such things before on his various crossings of the equator when sailing between England and Australia. But Hurley and Laseron were a hoot. They had done their Aboriginal routine early in their stay at Cape Denison, affirming the group's cultural and racial superiority in this place where they saw themselves as an outpost of civilisation on the edges of a wilderness, however uninhabited that wilderness might be.

Their routine helped Hurley and Laseron to feel one with the rest. Neither of these men had been to university. To make their fellows laugh brought them together, asserting a community of identity in opposition to blacks and women.

Later in the winter there had been further entertainments, in particular a comic mock-opera, 'The Washerwoman's Secret', with a plot as complicated as it was ludicrous. The washerwoman is ill and her daughter, Jemima, calls for the doctor whose fierce ministrations preclude all hope of recovery. In her dying moments the washerwoman reveals that her daughter is in reality a

Princess whom she has cared for 'since the revolution'. The rest of the 'opera' consists of the entanglements and machinations of various suitors for Jemima's hand. At the close villains are defeated and love is triumphant. Mawson noted that the opera concluded with a patriotic song.

'Old Joe' Laseron and 'Dad' McLean appeared as mother and daughter. Laseron, who is described by one of his fellows as 'a peculiar individual' and something of a 'sentimentalist', had already played a lubra, and now he took another female role. Mawson said McLean 'made a fine girl'. Behind a curtain which partitioned the hut, Laseron and McLean had transformed themselves under the gaze of some of their colleagues also preparing for the fun. With what secret thrill had they performed their cross-dressing, improvising breasts and stockings and skirts, feeling the peculiar mystique of women's dress and bodies. Who knows what orientations these men had?

When the players stepped beyond the curtain onto their improvised stage and heard the laughter and approval of their watching comrades, what misogyny was enjoined by the operation performed by hacksaw, hammer and chisel upon the body of Laseron in his role as mother? What frisson was sparked by the love-making scene between McLean and Hunter? For if such cross-dressing for the benefit of an all-male community acts to affirm their shared masculinity, it also covertly acknowledges the reality of same-sex desire. Herbert Dyce Murphy, wit and raconteur, also later known as a transvestite, watched from the audience, his laughter perhaps double as he kept his secret to himself.

There was also a cross-cultural element to the proceeding. The opera was conducted in French and 'a semi-foreign jumble'. Two of the characters had spoof German names: Dr Stakanhoiser and Count Spithoopenkoff. So the comic opera pointed up how ludicrous, not to say effeminate, the French and Germans are, while simultaneously titillating the 'real' Australian and British men.

28 January, 1913

MAWSON FOUGHT HIS WAY TO consciousness through clinging webs of dream. Food. It was always food. As he brushed the lingering images of unobtainable cakes and pastries from his mind, he became aware that there had been considerable drift in the night. He could tell by the cramped confines of his tent that he was hemmed in by snow. It sounded as if the wind had abated somewhat, but until he dug himself out he would not know for sure if it was going to be possible to make progress.

Mawson fought a rising sense of panic and despair. He felt as if he might lose his reason. The shrunken confine of his tent was like the narrow and obsessional space his mind had entered through these days of solitude. It was a kind of madness to be buried alive like this and to have no other voice or touch or sight of a living creature. All I've got, he declared to himself, are these limbs with their shrinking muscles and suppurating sores, my brains and my heart. Must dig myself out and walk on. Nothing else matters. Pain, panic, fear, don't matter. What matters is to act, to work. And to walk.

All his gear was either wet or frozen. He battled to get

his bandaged feet into the finnesko, then began tunnelling out on hands and knees. As he clawed and hacked his way out of the tent he allowed his mind to wander until he was walking through the bush to his first school at Rooty Hill with his brother. They had had to pass the blacks' camp every day, and often they would be taunted and chased by the scantily clad, painfully thin adolescent males of the mob. Mawson had hated the squalor of it all, and felt a chill of fear at what seemed like insuperable differences between them. He had always wanted to rush past this part of the journey, just as he was distressed now to get out and onward. He needed to escape from those parts of his mind that threatened his notions of civilisation and reason; the dark places that threatened who he was with thoughts of death, dissolution and abomination.

He made his mind move back to the Mawson house in Glebe, overlooking Black Wattle Bay, and the pleasant walk from there to Fort Street School. How harassed his mother had been in those days. But they had been grand for him and Will. School was fun and at the weekends there was camping and shooting in the nearby paddocks.

Faber est suae quisquae fortunae—'Each one is the maker of his own fortune'—the motto of his school came back to him. Never had it seemed more true than now, when he was so in need of all that he had ever learned. He considered that the training in manual and technical subjects and the emphasis on sport and physical culture provided by Fort Street had proved invaluable. Mawson remembered the crowded classrooms with their pictures of Burke and Wills on the wall and how they had learned about the explorers' feats. He aspired to their fame, but was determined not to share their fate.

He thought of the spreading arms of the huge Moreton Bay fig in the schoolyard. There were seats radiating from its trunk on which the boys had sat and yarned about their exploits on the games field, or planned their next trips in the bush. It was from here too that one could gaze with inexpressible feelings, almost mesmerised by the sight of the girls in their playground. From the figtree benches there was an uninterrupted view of these strange, frightening, winsome, ineffably exotic creatures. But then it would be up and off to games and girls would be forgotten again.

At last Mawson crawled into daylight to find that the wind had moderated, but snow was falling in large flakes from a dirty sky. The tent was buried to within a few inches of its apex, and there was no sight at all of his sledge and equipment. It was going to take hours of labour to liberate his gear. He began to dig, and as he did so his mind went back to Fort Street. What was it 'Boss' Turner had said about him when he left the school? Something about being a leader and organiser, and that if there was somewhere left to be explored it would be 'our Douglas' who would be sure to lead an expedition of discovery. At sixteen this had seemed the most romantic of possibilities. Now he tried not to think of the contrast between the romantic tales he'd been brought up on, and the misery of his present situation. He tried as he scrabbled to make the brute physical labour, the terrible isolation of his endeavour, the pain of his condition accord with the ideals of masculinity with which he had been inculcated.

What did it matter if he was alone? He was stronger for it, he told himself. There was none of the wrangling and tensions that had so exacerbated the horror of his walk back from the Magnetic Pole with Tweddy and Mackay. And would not his lone survival make a great story? Would it not capture the imaginations of generations to come? Here was more incentive. He must survive to make sure. He would succeed where his father had always somehow failed.

His father had always done his best, trying one venture after another. But somehow ill-luck always attended these endeavours. Mawson considered how he and Will had been fired to achievement by their mother. Social position would be re-gained through having a profession.

Mawson did not like to dwell upon the relationship between his parents. His father had often been absent from the domestic sphere, working long hours or travelling in search of the elusive fortune. Now, the necessity of his father's work in New Guinea provided sufficient rationale for the resolution of differences that had accumulated through the long years of his mother's disappointments.

Mawson considered that he had taken the best from both parents. He had inherited the adventurous part of his father that was impatient with the narrow confines of domesticity, but he had also learnt the value of meticulous attention to practicalities, a certain hardness in the face of the world's necessities, from his beleaguered mother. Surely they would both be proud of him. But it was always so difficult to tell what they felt.

Feelings. Best left alone.

Better to think of his father at last making some money and his mother comfortable in her little room in Sydney being looked after by Will. As for himself, he considered those memories of Fort Street as among his happiest. And as he laboured he saw the rather short, stocky figure of Boss Turner, belting out the school song. Turner had taught Mawson what it was to be an enthusiast. Discipline was the key to achievement. Any boy idling with hands in pockets was invited in a stentorian roar to a beating with a cane on the steps of the school; a public punishment that amply discouraged offenders.

Mawson knew he had to exercise discipline today. Impatience would not do, however anxious he was to get on. If only the weather would moderate, he was sure he could get home. But if it didn't ... Mawson turned away from this thought, and continued to shovel snow away from where he guessed his sledge was. As he continued the punishing labour, his mind became absorbed in the fight to ignore the physical pain of his endeavours; an interminable wrangle with himself to keep going so that he might win the ultimate prize: his life.

• • •

Fort Street School was established in 1849 as a model school to provide public education for all. Douglas and William Mawson matriculated from there in 1898. Special coaching, individual tuition, and extra teaching on Saturday mornings were available for aspirants to the university; all this cost only threepence a week compared with the twelve guineas a year for high schools. In the various celebrations which marked the school's jubilee, it was this relative egalitarianism of approach as well as nationalism that was celebrated in terms of high idealism by various politicians.

It is easy to gain an impression from this that Fort Street was somehow in the forefront of 1890s political radicalism. But it wasn't so. Empire loyalty was instilled by the cadet corps, and Empire Day was celebrated in all NSW schools. The school motto also pointed towards mainstream values of capitalism, rather than to any collegiate or collective values and articulates an individualistic ideal that finds expression in Mawson's attitudes to life.

Mawson's politics were mainstream conservative as represented by Fort Street School, and in all likelihood by his parents. It is difficult to imagine the Mawsons arriving from England in the 1880s suddenly involving themselves in the more radical Australian movements of the time which supported the ideas of unionism and republicanism. Given Mrs Mawson's ambitions for her sons, and Mr Mawson's adventures in capitalism, it is unlikely that they had much sympathy with working-class movements, much less a republicanism that would sever their emotional ties with 'home'. Mawson followed suit. He was what we might call an Empire Nationalist. In this he was joined by all his companions in Antarctica.

Mawson accepted the authority of the past both in science and in politics. Like the bourgeois heroes of the Empire's adventure fiction, he risked the unknown and endured the unspeakable in order to confirm traditions and impose inherited ideologies upon new lands. He had many of the characteristics often identified with quintessential Australian masculinity: his physique, initiative, daring, ability to improvise, interest in sports, limited appreciation of the arts. All these qualities are inherent in the bush myth and the myth of the Anzac soldier. But Mawson lacked one salient feature of these which has stopped him becoming a widely accepted cultural icon in the mould of Ned Kelly or the heroes of Henry Lawson and Banjo Paterson: he was not a working-class hero nor did he have any ambitions to be one.

• • •

The first hour's travelling had been a dispiriting slog in which he had again invoked Robert Service to remind himself that others had suffered similarly before him and survived. But the weary lines were losing their power to console him. The bad weather meant that he did not know if the ground he was covering meant anything; he could not be sure of his direction. And the cyclometer kept clogging with wet snow so that he could not be sure how far he had travelled.

If the weather did not break soon, the end of his story was not going to be a heroic tale of survival but a dismal failure—a final agony in this blank desert. He began to feel threatened by chaos, as if all the stories he kept telling himself were unravelling in the wind, so many useless words blown down the draughty cavern of emptiness that constituted the snow plain over which he crawled. The tide of panic could only be held back by prayer. In extremity he turned to Providence, which had protected him and walked with him so far. He prayed for a break in the weather. He yearned to be free of this perpetual greyness, to see the light and feel the warmth of the saving sun again.

After another bout of breathless pulling, his prayer was answered. Through cracked clouds a pale lemon sun shone forth. Mawson stopped to rest and to unclog the cyclometer again. The appearance of the sun also helped him to check his bearings. He moved on again with what he thought was good direction, under the blessing of a clearing sky.

The weather raised his spirits enormously and, fuelled with a little chocolate and some raisins, he plugged on, his sense of purpose and determination restored. He created all manner of welcomes for himself when he arrived back in Australia. Although he knew himself to be a shy man, and somebody who had to prepare a face to meet the world, nevertheless the thought of attention, of being hailed a national hero, had a beguiling attraction for him. If he survived, his reputation would be made, and whatever else life had in store for him there would be success to bolster him in adversity.

Welcome home from the City of Zero, he thought with a grin. The 'City of Zero' had appeared in a pantomime contrived by J. C. Williamson in honour of Australia's federation. *Australis* played in Sydney in December 1900 and was set in the year 2001 in an imaginary Australia. The pantomime concerned the adventures of a group of Australians who travel to the South Pole to rescue the queen of that land, Dione, from the wicked oppression of the wizard Azeemath. At the close of the show the land of ice has been incorporated into a 'great Empire of the South'. Mawson recalled the excitement of the show, the way it expressed the hope that the new nation might make its mark on the world stage by a daring act of adventure and imperial self-assertion. If the pantomime had been received with enthusiasm, he mused, how much more welcome his exploits in the real land of ice and snow would be.

And so Mawson pulled onwards through a long afternoon. But with every step his heart lightened. As if in answer to the terror and near despair of the morning, the balance swung back as he realised through the fine weather that he was descending from the plateau towards Commonwealth Bay, which he could discern by a patch of darker water-reflecting cloud on the northwestern horizon.

As he lay in his tent at the end of the day, he made his calculations again and felt that Aladdin's Cave was only about thirty-two miles distant. There was a real possibility now that if the weather remained kind to him he would reach the cave. It was still going to be a tight struggle for his food supplies were very low—about twenty chips of dried dog meat, half a pound of raisins and a few ounces of chocolate were left. But he could survive the three or four days needed to get to safety on these meagre rations. He knew he could. The sight of Commonwealth Bay had cheered him enormously. He went to sleep dismissing from his mind the possibility of bad weather.

29 January, 1913

IN HIS FIRST WAKING MOMENTS Mawson experienced a sense of wellbeing that somehow transcended his terrible physical condition and hunger. It was as if the mood of the previous evening had infected his dreams, though he could remember none. The night had given him an unusual respite from the torments of nightmares about food tantalisingly withheld or journeys miraculously nearing completion only to be thwarted by disaster. No, he had slept well, and busied himself to get organised for the day.

But this land of ultimate enigma had played its chameleon trick on him again. On gaining the outside of his tent, he was met with low drift and a strong breeze which he estimated must be blowing at about 45 mph. Clear skies and sunshine might never have existed. It was the familiar grey world that threatened to engulf him. But there were no options left. He had to travel on and trust to his navigational skills. He would use wind direction, the angle of the sastrugi, and the promise of Commonwealth Bay as he had seen it yesterday, as his aids. Chewing a piece of dog meat that had the consistency of hard leather, he packed his sledge and prepared for the day's effort.

Soon he was dragging his way forward again, hoping that his direction was good and that the weather might clear as it had done the day before. While the pain in his shoulders, legs and stomach was a constant reminder of his increasing frailty, he had put behind him the panic and near-despair he had experienced twenty-four hours earlier. His quest to save himself had regained its momentum.

He drove himself forward, knowing that the two pounds of food he had would not last forever. The question was how long could he make it last, and how far had he to go to get to Aladdin's Cave? To look at things from the worst possible point of view, it might take him another three or four days to get there. Which meant half a pound or less of food per day, and that much less energy to spend on the march. The obvious goal had to be to cover as much ground as he could as quickly as he could. At the end he might be able to march on will alone for a day or so but not much longer. The climate was too unforgiving to allow for any illusions of heroic survival without solid nourishment.

As these thoughts occupied his mind, Mawson glanced upwards, and ahead to his right saw something strange. Indistinctly raised against the horizon was some object or irregular shape that had a black colouring against the whites and greys of the landscape. He stopped for a moment and peered through the swirling drift, trying to make out more clearly what it was he could see. But he could not make it out, and so began to struggle excitedly towards it. Could it be the tent of some rescue party, or was it some odd trick of ice formation?

As he got closer to it, Mawson could discern that he

was staring at a man-made cairn that was surmounted by what looked like a black bag. With thumping heart and straining muscles he pulled towards it in a paroxysm of impatience. As he pulled alongside and clumsily disentangled himself from his harness, his hopes soared. The bag on top of the cairn looked like a food bag. He rushed to confirm this. Joy and relief flooded through his being as he realised it was. In an instant he felt that he had moved from the valley of the shadow of death, back into life. There would be no more anxiety about food.

But there was no time to dwell on this miracle as Mawson seized a tin that had been left next to the food bag. He wrestled with it, clumsy in his impatience to find what further treasures were here. Inside the tin was a note dated this same day and signed by Alfie Hodgeman, Hurley, and 'Dad' McLean from the party at Commonwealth Bay. As he read the date and the message, Mawson's eyes immediately peered into the drift, desperately hoping that he would see the figures of his colleagues somewhere in the distance through the seething flakes of snow. But they had left that morning, only a few hours earlier.

Mawson's awful disappointment was balanced by an extraordinary sense of gratitude to Providence. It was a miracle, he thought, that in all this icy waste he had stumbled upon this cairn that would save his life. There was more good news. The note said that he was only twenty-one miles S60°E from Aladdin's Cave—much closer than he had calculated. There was another cairn at fourteen miles S60°E with further food supplies. And in the bag provided there were biscuits, pemmican, ground biscuits, tea, chocolate and butter. The note also conveyed the news that the *Aurora* was at Commonwealth Bay, that wireless messages had been received, and that all parties were safe.

It was almost too much to cope with at once. In his excitement Mawson had to keep reading the note over and over again so that he could take everything in. Amundsen, it said, had reached the South Pole in December 1911 and stayed there for three days. A supporting party had left Scott one hundred and fifty miles from the Pole in the same month. It seemed that Scott had been obliged to stay in Antarctica for a further year.

Thinking of these things, Mawson emptied the food bag, scattering packages all around him on the ground. Here was a feast. The prospect lay before him like a picture of some medieval banquet where there are more courses than can be good for the digestion, however tempting they are to the eyes and the nose. This thought reminded Mawson that he must be careful not to overdo his intake. He had memories of the gorging they had indulged in after the march to the Magnetic South Pole and of the sickness and diarrhoea they suffered after their binge. Care was needed not to make himself ill.

As he nibbled luxuriously on a piece of chocolate and packed the other items back into their bag, nothing could stop his exultation. The *Aurora* was still at Commonwealth Bay. He must hurry forward. He did not wish to emulate Scott by having to spend another winter in this desolate place. Scott would be labouring under massive disappointment that Amundsen had got there first. Despite his own struggle, which was not yet over, Mawson felt vindicated that he had chosen not to go with Scott. To be first to the Pole was a scientifically meaningless gesture, but a gesture nevertheless. To be *second* at the Pole was hardly the stuff of national or imperial celebration. And to be beaten there by Norwegians. Scott would feel humiliated.

Mawson's reflections were interrupted by the necessity to move forward. There was still the problem of navigation to deal with. Despite uncertainties about his direction, the rest of the day passed quickly, his mind uplifted by the thought that he could have a hearty meal tonight. Surely nothing could happen now to prevent his reaching Aladdin's Cave and safety. He might have to endure more pain and discomfort, and suffer more of the loneliness which had made him converse with himself, but he would survive.

29 January –
8 February, 1913

Low drift and 45mph breeze blowing. Kept course N45°W for 5m then saw 300 yds to N of me a cairn with black cloth on it. I went to it and found food and note from McLean, Hurley, Hodgeman— they had left the same morning. 21mS60°E from Aladdin's Cave. I went ½ mile on that course, then thinking meant [sic] N30°W, I went 1½m on that, then turned to N60°W and did total of 8 miles from cairn.

Heavy work putting up tent that evening. Drift hazed everything, so could not well see landscape, etc but apparently ocean coming into view to N with icebergs. It almost seems to me that I am coming too far W—did McLean give the correct bearings? If clear day I might see them—I must be near the cairn mentioned at 14m but cannot see.

Am now on ice and half ice and falling every few yards on account of heavy side wind, so have to camp before I should otherwise. Distance for day 13 miles.

It must be about 13½ miles to Aladdin's cave. If

calm day tomorrow I can do it, not otherwise as have no crampons and keep falling.

It is a great joy to have plenty of food but must see that don't overload or disaster may result. What a pity I did not catch McLean's party this morning.

Mawson put aside his diary and pencil and tried to compose himself for sleep, but he felt a terrible restlessness and impatience. He was simultaneously elated and frustrated, feelings which together with the richness of his meal induced an unpleasant whirling sensation in his gut. If only tomorrow the weather would be kind. If only he'd seen his comrades this morning. But the drift in this forsaken landscape made everything impossible. Mawson's mind kept playing with the possibilities of distances and directions until he fell into a restless doze, trying to imagine his ideal of clarity—a condition of mind and space governed by strict lines and rules, the unambiguous congruities of science.

The morning brought more wind and drift. Mawson, after another luxurious meal, busied himself with improvising crampons from bits and pieces of his meagre equipment. In the afternoon the wind declined and Mawson set off, slipping and sliding like a drunken ice-skater, stumbling, sometimes falling as his crampons broke into pieces under the strain. He also felt on more than one occasion the sickening feeling of the sledge breaking through the lid of crevasses as if to remind him that all was not yet safe; that he might be robbed of his life yet with so few miles still left to travel.

After five and a quarter miles of intermittent progress,

Mawson had to stop. His crampons were destroyed and without improvising some more, further progress was impossible. But this afternoon he had seen the open sea and the great iceberg known as the pianoforte waiting for some giant impresario to play. Mawson had found the sight of the sea inspiring, speaking as it did of the possibility of escape, an exit from the enclosed world of white, ice, drift; the world inside his head.

More food. The comfort was beyond words. But then there was the problem of crampons to occupy his mind. He took apart the theodolite box, removed nails and screws out of the cyclometer and cooker box, and sat fashioning these rough materials into the footwear that would get him over the ice. As he did so he discovered he was talking to himself. The secret is to enjoy this. To say, look how good, how capable I am, look at my strengths. Only misfortune can beat me. But I won't league with fortune in my own defeat. Not far to go now. Not long to go now. Won't do to think too much about others. They aren't here. Can't help. Only me. Only my efforts. Yes, enjoy the loneliness. Nobody to interfere. Nobody to say stupid things. Nobody to want responses. Only my needs, only my efforts.

On 1 February 1913, after further struggle and difficulties with crampons, Mawson arrived at a cairn he had seen two days earlier and found it to be Aladdin's Cave. Never had it seemed to deserve its name so richly. Mawson's joy at entering this icy grotto was like a child's at the magic of pantomime. Inside, someone had left fresh fruit and further stores of food. And he was only five and a quarter miles away from the hut. 'Great joy and thanksgiving,' Mawson pencilled into his diary, snug beyond the wind, in the soft light of the ice cave. He felt safe and secure, burrowed beyond the terrors outside. But still he knew he had to act. The crampons had to be improved. For the road downhill to the hut was all ice, and he could not travel without some purchase on it. His colleagues had not thought to leave spare crampons for him.

The next days provided Mawson with exquisite mental torture. The wind and drift continued too strongly for him to risk moving. So he spent the time knocking apart old benzine cases and scouring the immediate environs of the cave to find anything that might help with the problem of the crampons. He found some old dog harness and more screws and nails, so he worked away with his bonza knife, trying to make the intractable materials into the means of his descent to the hut.

Mawson hated this last delay. It strained his impatience and harassed his brain. Useless questions would revolve over and over. Would his colleagues have waited for him, if the ship had been and gone? What would become of him, if they'd left him to his fate—a literal Robinson Crusoe on this least fertile, most hostile of islands in the ice? Would he be able to survive for another year alone without succumbing to madness, without destroying himself?

Mawson tried to keep such harrowing thoughts at bay. He repeated his stories to himself. I am strong. I will survive. I am the leader. But as the weary days and nights dragged on he felt sometimes as if he no longer knew who he was or what he was doing. He began to feel lost in a mental blizzard where nothing could be defined any longer. Everything was losing its shape and place. It was as if he was losing himself, sitting alone in the ice cave like a man from some distant age surrounded by savagery and the implacable, intolerable, carelessness of nature.

I must not succumb to this. I must hang on. The weather must change. And I have food. I've come so far. I can't be beaten now. Must concentrate on what is. The reality. Only five and a quarter miles to go. Imagine the ice road is like Glebe Point Road. Normal. Nothing different, nothing terrifying. And at the bottom of the hill are men who know Fort Street and Sydney Harbour, all the old familiar places, palm trees and sunshine. It is all within the grasp of my mind. I must not let go.

And so a week passed. A week in which every day Mawson got up, hoping that the wind would have died sufficiently to let him walk upright, to let him walk with confidence towards the hut. But it was not to be. And as he waited he realised his body was beginning to succumb to scurvy or some such problem. His nose kept bleeding, but not with the rich claret of the football field. Rather a sort of thin, pale red substance was constantly dribbling onto his lips, and he could feel the warm salt on the back of his throat. The ends of his fingers too were bleeding and suppurating. Mawson tried variations on his diet. But there was little that would relieve his worries. To die within such a short distance of the hut seemed like the most awful irony of fate. But he would not die. He refused.

On 8 February Mawson broke out of the cave and with a giddy feeling of excitement realised that the wind had dropped during the night. He could set off. He stood for a few moments to check that he wasn't dreaming or imagining the more benign conditions. It was difficult to believe after so long a wait.

He busied himself to get under way and soon was careering down the icy slope. Sometimes on all fours, sometimes in an ungainly stumble, he passed the miles with mounting anticipation. Again, he rehearsed his litany of anxiety. Would anyone be there? Would the ship have left? Would he find himself safe at the hut, but intolerably alone? He felt terror at the thought of the last eventuality. He could not stand such an end to the journey. He felt the tug of madness threatening.

About a mile away from the hut, he saw the anchorage opening up before his eyes. It was empty. No sign of the ship. He forced himself forward. Soon, soon he would see the harbour and the hut. Yes, yes, there surely at the harbour, were they not figures, black specks moving against the grey and white? He stopped to peer breathlessly into the distance, unwilling to believe his eyes; afraid of the disappointment should he be deceived.

Then he waved. And waved and waved with increasingly frantic gestures. He saw the answering salute and nearly collapsed to his knees with gratitude. He began to stumble downwards and as he did so saw the distant figures begin to run towards him. Because the road was steep it was several moments before Mawson saw the first of his comrades approaching. He wondered who it would be. At fifty yards he could not tell, but soon he saw Bickerton approaching with a look of awe and wonder on his face. Soon they were in each other's arms and Bage, Madigan, McLean and Hodgeman all arrived, and there was such noise and confusion as they all wanted to hear his story and tell theirs.

Supported now by strong arms, Mawson was led down to the hut, and he told his story as simply and quickly as he could, and as he did so he noticed there were tears in the eyes of the other men. But there were no tears in Mawson's eyes. He felt overcome with a feeling of relief and gratitude. A letting go of the strain. But he could not weep. For was he not a man of will and strength? The leader. The survivor.

epilogue

WHERE DOES ANYTHING begin or end? The journeys, the mythologies go on and on, endlessly written and rewritten, painted, photographed, sculpted. I'm standing in Waterloo Place, London, staring at the statue of Scott modelled by Kathleen—one of several she made of her husband in the years immediately following his death. Under a grey, moody sky he stands there curiously removed from the bustle of the London streets, the everlasting sound of traffic and the wail of sirens, peering into some invisible distance. And I think of Kathleen moulding him.

I see her in her workroom. Pale lemon peels of sunlight are strewn across the wooden floor. Dust motes dance as if excited by the first delights of spring. Kathleen is busy and grateful for these mornings when, with Peter playing at his wooden toys nearby, she can lose herself in her work. And what work it is, her hands at their sensuous task, re-creating from the wet clay the limbs, the torso, the head and features of her husband. Kathleen worked from Hurley's photographs aided by her memory and imagination. She was creating the perfect man in heroic pose, standing resolute, his gaze straight as if to the horizon, suggestive of the visionary. His right arm is outstretched holding a ski stock at a forward sloping angle, which makes it look like a staff. This and the medieval resonances of his garb, particularly his cowl-like hood, give the finished piece a suggestion of Christian asceticism as well as one of determined resolution in the quest.

As Kathleen caresses Scott's thighs into being, concentrating hard on the folds and fall of his trousers, she finds it difficult not to think of the rushed confusion of their embraces. The warmth of their tangled limbs contrasts starkly with the cold clay. Yet here, now, she is in control. She exults in her ability to transform the final days of suffering into this lasting memorial to greatness. Now she can possess him entirely and make him whole, without flaw. Her hands wet with the malleable clay, she feels at one with him in the rhythm of her making; now there is no distinction between them. As she works, Kathleen wonders if all human transaction is like this—always seeing in others what we want to see—split and refracted images of ourselves.

Kathleen had received news of Scott's death on board ship as she steamed towards New Zealand, expecting to be reunited with her husband after an absence of more than two years. She dealt with the news in the same way that Scott had dealt with his own demise, by transmuting it into a metaphysical victory. She confided to her diary, 'My god is godly. I need not touch him to know that. Let me maintain my high, adoring exultation, and not let the contamination of sorrow touch me. Within I shall be exultant. My god is glorious and could never become less so.' A few days later she is writing that Scott has made her 'twice the man she was', that he is her 'motive power' and always will remain so. Kathleen was already transmuting her husband into an image, an icon of an ideally heroic figure which she herself aspired to be.

I take a train from London to Cambridge where I have arranged to visit the Scott Polar Research Institute for the second time. It was funded with money left over from the public appeal set up in the wake of the disaster to provide for the families of the deceased in consonance with Scott's final plea. In death as in life, class differences were observed. Kathleen Scott received £8,500 and a pension of £300 a year; Edgar Evans's widow received £1,500 and £48 a year. Money from the fund was also used to finance the publication of the official reports of the expedition and several memorials, while £10,000 went to an endowment fund for the Research Institute, which is now a part of Cambridge University. Here work about both North and South Polar matters is carried on by devoted librarians, teachers and researchers; the place gives to Scott's memory an aura of scientific respectability at the heart of the British academic establishment.

On my first visit I was met with amiable suspicion. It is made clear to me in the friendliest way possible, that if I were seriously interested in Polar matters I would be researching earlier nineteenth-century expeditions rather than that of Scott. Scott has been 'done'. And, of course, there is the fear that I might join the muck-raking tradition inaugurated in 1979 by Roland Huntford's iconoclastic and judgmental approach, which caused so much furore with Kathleen Scott's descendants. 'I hope you're not writing a novel', is a phrase I remember hearing, with the word *novel* pronounced as if it were a particularly filthy form of contamination.

Despite the air of protectiveness and the advice, amiability is preserved, and I'm here again to look at some of the documents relating to the aftermath of the expedition. There is a rich haul, including several albums of press clippings. I find myself wondering who made the clippings, since Kathleen was still in ignorance when the news broke in London and when a memorial service was held for Scott in St Paul's Cathedral a few days later.

On the morning of 11 February 1913, England woke up to sombre headlines proclaiming the death of Scott and the Polar party in all its major dailies. Here was the end of a story that the public had followed avidly through the newspapers from the beginning. What had been trumpeted in 1910 as 'a gallant effort to uphold the supremacy of Britain in the Antarctic', had now turned into a great 'sacrifice' for the Empire.

The newspapers indulged in paroxysms of high-flown rhetoric and led a public outpouring of grief not entirely unlike that witnessed for the death of Diana, Princess of Wales. Many papers on the days following ran special pictorial editions with black margins, depicting the dead men and the bereaved, including Peter Scott and the three small children of Petty Officer Evans. Scott's 'Message to the Public' and last words were widely circulated.

The idea of martyrdom for the Empire was much to the taste of both press and public. The editorials concentrated on 'glory' and 'heroism'. Under the headline 'Victory and Death', the *Daily Express* trumpeted: 'Weep not for the Dead, Captain Scott and his companions have not died in vain, nor can their glory perish.' The *Daily Telegraph* agreed: 'Scott, dying in the prime of his manhood, is added to the country's imperishable roll of heroes.' The *Daily Mail* saw the deaths as an advertisement to other nations: 'Foreign opinion is a mirror in which we see ourselves reflected, and in this case the reflection is one that shines with glory.'

Several papers asked rhetorically whether the benefits were worth the costs, only to affirm the meaning of the deaths in the manner of the *Pall Mall Gazette*: 'It is a white, and not a black mourning that we wear for these gallant souls ... So long as there is any spiritual meaning in life, the heart will teach us that to exist in a safe, organised routine is not the highest fulfilment of human destiny. It is ever the hero who carries the true lantern of our pilgrimage.'

Scott's own prose was having its desired effect. The reasons for the 'tragedy' were canvassed but overwhelmed by patriotic, imperialist and religious sentiment. A plethora of biographical sketches hailed Scott as a hero and constructed from his story an exemplary life, so inaugurating a tradition that was to continue for the next seventy years.

Militarism was never far away from the various accounts of Scott's death and the meanings generated from it. The language of victory and conquest was habitually used. Yet it is difficult now to see what or whom had been 'conquered' and wherein lay the victory. But on the brink of World War I no one was going to question this too closely. The death of Scott and his companions was to be marshalled to show others how to sacrifice themselves for God, King, Country and Empire. Scott was being made into myth.

As I read the papers I thought of Mawson. On 11 February it was his third day back at the hut. He was still hoping that the *Aurora* would return and be able to take them home. By 14 February, the day of the memorial service for Scott, it must have been dawning that they were marooned for another year. It is a terrible thought.

Mawson knew that he had been 'shaken somewhat to pieces'. He had to keep sitting down. His legs are swollen, and periodically he feels faint and dizzy. Yet as the others go about their tasks inside and outside the hut, Mawson finds it difficult to stay still. He staggers into the kitchen area to stand beside Madigan as he prepares lunch, touching the latter's sleeve as he cuts the bread; he follows Bage and Hodgeman as they begin to arrange the stores of seal and penguin carcasses for another winter. Not that Mawson has anything to say to these men or that he can help them. Rather, he gains comfort from their close physical proximity.

When sitting alone he begins to feel a rising panic, as if the others might disappear, as if he might awaken and find himself alone again in that loathsome makeshift tent. In the company of others he can focus his mind on the present and reassure himself of the reality of the hut. Sitting alone, somehow his mind refuses to concentrate, and he discovers himself in a mental drift that threatens to disorientate, even engulf him. He finds himself crawling over the ice again, haunted by the ghosts of Ninnis and Mertz, whose voices he hears plainly though they issue from crevasses and across gale-swept wastes.

Mawson's physical symptoms, however, anchor him

in a different reality. He has difficulty assimilating any food. Frequent struggles to the luxury of the can remind him that he is truly back at the hut, that he has shelter, that food is to hand. Yet physical deterioration frightens him as well. He is emaciated. There are scabby sores all over his face and boils on his body. His legs and feet are in a terrible way. And the loss of so much hair preys on his mind. He does not want to be bald.

There are letters from home. Will has written telling of his father's death. Mawson can hardly comprehend this further loss. He feels bludgeoned. The initial elation of his arrival at the hut will not return. Mawson feels alternately numbed or violently agitated; it is as if his soul is frostbitten. There is no rest. He sleeps, only to be pursued by the same spectres that harry his waking nightmares: his father and mother, Will and Paquita join Ninnis and Mertz in a grotesque scurry over the ice to catch him, while he crawls desperately forward without making ground, until suddenly he is plummeted into wakefulness as he falls through the lid of his crevassed dream.

Then they receive a message telling of the death of Scott and his companions. Mawson is shocked and saddened again. He could so easily have been with them. Would he have succumbed in the same way? Two expeditions, seven men dead. And he the survivor. Why? How? Providence seems inadequate to explain the miracle. He keeps returning to the message, trying to take in the news. But words and images spiral away in a maelstrom of thoughts and feelings, like so many snowflakes blown into whirlies by the storm. Nothing in his mind will settle.

• • •

A sudden bell's ringing summons me back from Antarctica to St Paul's. But it is the ship's bell of the *Terra Nova* rung at the Scott Polar Research Institute to signal morning coffee and afternoon tea. I decide to keep reading. Time is short and I've been dreaming. I turn back to the newspapers with their photographs of the crowds outside the cathedral and their further gush of by-now familiar rhetoric. Had Scott survived, he would have been known as the man who was second at the Pole, a minor celebrity perhaps, but little more. As it is, he has become one of the great martyrs of Empire. At the same time as the memorial service, the Education Committee of the London County Council decided that Scott's story, his message to the public and final words would be read to school children in all elementary schools in London. Many of those in provincial cities followed suit. The committee felt that there could be no greater example, no finer lesson for the children, than that expressed by the story of the last days of Captain Scott. Their decision demonstrates the very fine line between education and propaganda.

As I work through the rest of the memorabilia—postcards, poems, lecture programmes, Ponting's film—I am astonished at the insistence with which the key ideas are reiterated. I see now how the outbreak of World War I ensured that the process begun by the newspapers would be carried on by other means. And on and on. Scott's story was made into propaganda for the war effort. Ponting's film was used at the base camp in Etaples to show thousands of troops how to die like 'English gentlemen'. Kathleen's sculptures, the one in Waterloo Place and another one in St Paul's Cathedral, were unveiled in 1915 and 1916 respectively, with appropriate speeches by politicians who did not fail to draw the comparison between Scott's death and the sacrifice demanded by the nation of thousands upon thousands of young men in the great blood-letting of Ypres and the Somme. And in case the working classes had missed the significance of all this, they could, upon opening their favourite packet of Woodbines or Players, find there on cigarette cards images of Scott or Oates, Bowers or Wilson to remind them of their duty. It seemed to occur to no one that being machine-gunned, shelled, or bayoneted was hardly comparable to going to sleep in the snow.

On my last day in Cambridge, I stand outside the Institute looking at the sculpture in the front garden. It is another of Kathleen's works made during the war in black marble and shows a young male nude in an attitude of Christ-like sacrifice. The inscription on the pedestal reads: 'These had most to give.' It makes me feel angry in its complacent deceit. As I walk away I'm thinking of my grandfather and all the millions of working-class men like him who were suborned by such images to give away their lives or parts of their lives or parts of their bodies in the squalid horror of the trenches, and of all the bereaved for whom such images were supposed to provide comfort.

• • •

I leave England for Adelaide and the punching heat of summer. Mawson's archive has been moved out of the Barr Smith Library on the university's main campus to Urbrae House in a more suburban setting. The atmosphere of both places is much more open, friendly and relaxed than Cambridge. From the beginning, at the Barr Smith, the staff have been enthusiastic to help and discuss my project in a non-intrusive way. 'I hope you're not just writing hero-worship hagiography,' said one librarian. 'I hope so, too,' I replied. At Urbrae House, the tradition is continued. The curator is an Antarctic enthusiast; we have helpful discussions over lunch. There is never any problem with access to materials.

As I research the aftermath of Mawson's expedition, I begin to realise just how far the publicity surrounding Scott cast Mawson's achievements into shadow. The timing was terrible for Mawson also. While the apotheosis of Scott was going on in 1913, Mawson was enduring another terrible winter in Antarctica. The wireless mast was up and working for some of the time, but only short messages could be sent, and his news could not compete with that of Scott.

But worse than all this, as the dark, tempestuous winter wore on, was the descent into madness of the wireless operator, Jeffryes. Seven men in the hut; Mawson attempting to recover from his trek; Jeffryes exhibiting the symptoms of psychosis. No wonder, as he sits in his alcove, cradling his head in his arms, Mawson feels fury mounting as he regards the confusion of books and papers on his desk. Day after day it is the same. He imagines his brain exploding into fragments. All the frustrations of the day boil over as he hurls a book he's been reading against the wall, before sitting down breathless and close to defeat. His life is buried under heaps of paper that are somehow far worse than the falls of snow that had threatened to bury him on his march home. He has been struggling to write all day, working on his account of the expedition, and as so often happens he is now over-tired and overwrought. He feels surrounded and threatened by insanity. He experiences a tightness in his skull and in his gut as if his body contains a massive pressure that cannot be endured much longer.

Every day his mind circles the same territory, burdened by thoughts of his father's death, his mother ill in Sydney, Paquita waiting and anxious, the debts the expedition was incurring, the difficulties of publicising his efforts, the obligation to assert himself as leader, the petty irritations and disappointments provided by his fellows; and now, to crown all, there is Jeffryes's madness to contend with. On the trek home at least there had been a goal each day, and on most there was some progress bought at the cost of horrific exertion. But here, now, he is imprisoned. There is no escape from the mind's bleak wastes, physically constrained by the hut and the gales that battered it.

In early July, in the middle of the Polar night, Jeffryes had shown the first signs of madness. He began to offer physical violence to his fellows on the grounds that they were plotting against him. There were confessions of how he had 'gone the pace' as a young man and suffered venereal disease; he had begged Mawson to allow McLean to administer poison. Mawson had tried all manner of threat and persuasion to control the aberrant behaviour. Nothing worked. In his better moments, Jeffryes relapsed into almost total lassitude, lying on his bunk all day, neglecting his personal hygiene to such an extent that tins of pump-ship and 'rears' were left in the hut. At his worst he raved about the plot against him, how others were trying to murder him, how he would have them all arrested when they went home. That Jeffryes was the wireless operator only exacerbated the problems; they could never be sure what he was transmitting and often could not cajole him into listening for incoming messages. On one occasion they realised that he had sent a message to Macquarie Island alleging that there was a plot afoot to murder him.

Every interaction with Jeffryes left Mawson feeling unsettled himself. The curiously logical way in which Jeffryes articulated his delusions aroused in Mawson a kind of panic, a suspicion that it was perhaps himself going mad and not Jeffryes at all. And Mawson had to league with the others to make sure Jeffryes was watched at all times. This only increased the madman's paranoia. Jeffryes's accusations that Mawson was plotting against him had to be denied. As he did so, however, Mawson knew he was not being straightforward. In this way he was dragged into Jeffryes's circle of paranoia and delusion.

At such times as this, Mawson tries to cling to the idea of another life that waits beyond the physical and mental blizzards. He has Paquita's letters, her faithful words. But he is anxious. He has been able to send her so few wireless messages, but now the wireless mast is down again, it isn't possible to send, and he feels guilty that he has not sent more messages earlier, and frustrated that the silence between them is now once more complete. When the weather permits he will get all hands to work on the mast, but at the moment it is impossible. For now all he can do is to exert himself with push-ups and sit-ups in the forlorn hope that this will help him to sleep and relieve him of the nightmares that haunt both his sleeping and his waking.

As I turn back to read Paquita's letters of this second year again, it becomes clear that Mawson was right to be worried about her. As the year dragged on, she became more impatient and insecure about their future. The fact that she had received so few messages from him began to irritate her. She wonders if he is frozen in heart and asks if a person can remain in such cold and lonely regions and still love warmly? She longs for words from him and pleads, 'You will not go again will you?' Later, Paquita feels as if she is writing to a wall and describes the everlasting silence as almost unbearable. She doesn't want to doubt him but wonders if she will satisfy him and wants to be convinced that 'all will go well with us and with our love'. She longs for their meeting but in a faint way dreads it. And in the last letter she wrote before the *Aurora* sailed, she says, 'It is no good depending on you ... If you had wanted to wire you could have ... One sided correspondence is the limit.'

Both Paquita and Mawson had to bear the loneliness of their hopes and fears, knowing that only their reunion could decide the future, if any, of their relationship.

I turn to the scrapbooks of press cuttings which tell the story of Mawson's arrival in Australia, his marriage and his subsequent journey to England. Although, of course, there is some celebration of his achievement, there is in the Australian press a tendency to ask much tougher questions about what actually had been achieved than were asked about Scott, and a questioning on occasion of whether too much was made of 'Antarctic heroes' at the expense of the ordinary battlers of everyday life. In both the British and Australian papers there is a continual comparison in which Mawson's expedition is implicitly or explicitly seen as secondary to the 'glorious' story of Scott's demise.

Mawson arrived back in Adelaide on 26 February 1914. The edited version of Scott's diaries had been published to widespread acclaim the previous November. Ponting's film of the Scott expedition had been running in London and Australia and Hurley's pictures of the Mawson expedition brought back the previous year had been running at a loss. For all the civic receptions in honour of Mawson held in Adelaide, Melbourne, Sydney, he must have known that he faced an uphill battle to push his own endeavours to the forefront.

• • •

Between days of research there are snippets of conversation. My wife's Aunt Margaret, now a retired school headmistress, remembers saying 'Good Morning' to Sir Douglas when she was a student at the university. 'He always seemed gentle and courteous. He had a lovely smile.' Another friend confides that her father knew Mawson well and she has childhood memories of the great man—again the impression is of a large but benign presence. He must have mellowed. There is evidence in the archive that Mawson was not slow as a young man to become embroiled in university politics and to pursue the interests of his own career with a certain avidity.

There are also stories and rumours of much later discontent and difficulty in the Geology department at Adelaide University while he was professor there. There are stories of muddled administration. He was no saint. How could he be? Nobody with that tenacious will, energy and ambition was going to be an easy person.

But in 1914 it was an extraordinary time for Mawson and so much was still before him. The relief of getting home, the reunion with Paquita, the civic duties and preparations for his wedding were all taking place in a context of anxiety about money and publicity. And then there were the aftershocks of his ordeal. I think about the night of 30 March, the night before he was married. There had been a banquet in his honour—another public display. But alone in his hotel room that night in Melbourne what restless excitement, what thoughts crowded into his mind?

Mawson paces up and down his hotel room perspiring profusely. Ever since his arrival home he has felt the oppression of heat. He cannot sleep. The evening has passed in a blur of fine food and wine and brilliant speeches made in his honour. As he has done on most days since his return, he has donned his public mask of charm, consideration and modesty. But now alone at 3a.m. he treads the polished floorboards, watching their interlocking lines. He feels as if he is treading between his past and future. The idea of the future that so sustained him on those icy days alone is now upon him. It hardly resembles his dream. It is not that he feels disappointed, more that he cannot tell with any certainty what the future might hold. Tomorrow he will be married. Then to England. Once he arrives in London he must settle to the book of the expedition, and make sure the film is mounted. These will bring the money in.

But before that there is the wedding to be endured. It will be a huge occasion. The papers have been making much of it, and it is without doubt going to be one of *the* society weddings of the year. He dreads it.

Thinking this makes him wish he was already aboard ship steaming towards England. He is immensely proud of Paquita, and proud to be her husband. What Mawson hates is the actual performance of his public duties—being stared at, small-talk, the obligation to be nice and sociable with people with whom he has nothing whatsoever in common. At some level it bores him. He wants to lose himself in meaningful action. He cannot see when this will be possible again. It is going to be a desperate scramble for money and publicity.

And there are other obligations to fulfil in Europe. He will have to see the families of Ninnis and Mertz. Mertz and Ninnis, Ninnis and Mertz. The names clang and boom in his head. Always there, always haunting. And the correspondence he's had so far with the Mertz family has been exceedingly difficult. One of the Mertz brothers has written in aggressive and aggrieved tones and is clearly after some financial compensation. Mawson has therefore been obliged to be strict in his dealings with them. And then there have been the further troubles with Jeffryes, who kept sending Mawson crazed letters from Royal Park mental hospital full of religious delusions and fulminations against the 'damnable hypocrisy of freemasonry and its injustices'. On top of this he has also had an acrimonious correspondence with Jeffryes's family.

The spectres of death and madness do not leave Mawson's mind for long. Every day, for one reason or another, he is forced in fragmentary images to relive the deaths of his comrades, the extremity of his experiences in crevasse and blizzard. He asks himself over and over again if their deaths could have been avoided. He has to reassure himself that he is not to blame.

And so he paces on, waiting for the dawn, as he so often waited in his makeshift tent—waiting for his life to start again. It is perhaps as well that Mawson could not see what the next few years would bring. He could not foretell how his book and film were destined to fail, caught forever in the shadow cast by the Scott expedition; and how the outbreak of war would ensure the life and pertinence of the Scott legend while simultaneously destroying the potential of his own.

Mawson's book, published at the beginning of 1915, sold poorly in England and Australia. The public was too engrossed in the war to be interested in another Antarctic book, not to say one without the sensational appeal of Scott's and poorly written in places, with a narrative uneasily poised between adventure story and scientific description. Mawson's film shared a similar fate. It opened in London on 20 April 1915 and ran at a loss for only thirteen days; days in which the British and Australian public had plenty of other things on their minds. Two days later at Ypres the Germans first used poison gas with some success, the allied troops at this stage having not been equipped with gas masks. The following day, another English legend was born with the death of Rupert Brooke on Skyros. He was en route to play his part in the ill-fated Dardanelles campaign which began on Sunday, 25 April—Anzac Day.

Both in England and in Australia, the contribution of Mawson's expedition to Australian renown was completely submerged by the 'Thrilling Deeds of Heroism' performed by the Australian troops at Gallipoli and proclaimed in the British newspapers with banner headlines. The Anzac landings appealed to the Australian taste for working-class heroes, ordinary blokes making good; Mawson could never quite enter into this category. As if all this were not enough, on 5 May, an English passenger liner, the *Lusitania*, was sunk by a German U-boat with massive loss of life, causing great consternation among the public. With it Mawson's film also sank. On 5 May it closed at the Alhambra theatre.

But this is not how we should leave Mawson. Nor should we look so far forward to the time in which he is known only as a figure on a recent banknote, the subject of advertisements for financial products, the man of achievement converted to the hard currency of cash exchange.

Let us return to his pacing vigil, as he awaits the bright morning that will bring him to his bride. Encumbered though he is with doubts and troubles, he tries nevertheless to look forward with hope to the uncertain future, just as he did on so many seemingly hopeless days on his lonely journey home.

Before I leave Adelaide, I walk along North Terrace one more time. Someone has told me that once, at one of the entrances, there was a bust of Sir Douglas Mawson, but that a car crashed into it and destroyed it. Perhaps it is better so. Better to let the lives of the dead inhabit our minds and imagination in a dynamic and shifting way, rather than to be cast in stone. As I'm thinking of this, I see out of the corner of my eye a tall old man in a dark suit. He is balding with white hair. His eyes are ice-blue, his jaw square. As he strides past me I gasp. I think I have seen Mawson. As I watch the figure stride away from me, I smile to myself and walk on.

references

In addition to unpublished diaries and letters held at the Scott Polar Research Institute and the Mawson archive at the University of Adelaide, the following books and journals have been of particular help.

Scott:

Scott's Last Expedition: The Personal Journals of Captain R.F. Scott, R.N., C.V.O., On his Journey to the South Pole, (London: John Murray, 1923); Edward Wilson, *Diary of the Terra Nova Expedition 1910–1912*, ed. H.R. King, (New York: Humanities Press, 1972); *The Norwegian With Scott: Tryggve Gran's Antarctic Diary 1910–1913*, ed. G. Hattersley-Smith, (London: National Maritime Museum, HMSO, 1984); *Under Scott's Command: Lashly's Antarctic Diaries*, ed. A.R. Ellis, (New York: Taplinger, 1969); *Silas: The Antarctic Diaries and Memoir of Charles S. Wright*, ed. C. Bull and P.F. Wright, (Ohio: Ohio State University Press, 1993); A. Cherry-Garrard, *The Worst Journey in the*

World, (Harmondsworth: Penguin, 1922); Admiral Lord Mountevans, *South With Scott*, (London: Collins, 1921); H. Ponting, *The Great White South*, (London: Duckworth, 1922); F. Debenham, *In the Antarctic: Stories of Scott's Last Expedition*, (London: John Murray, 1952); S. Gwynn, *Captain Scott*, (London: The Bodley Head, 1929); Reginald Pound, *Scott of the Antarctic*, (London: Cassell, 1966); E. Huxley, *Scott of the Antarctic*, (London: Weidenfeld and Nicholson, 1977); D. Thomson, *Scott's Men*, (London: Allen Lane, 1977); R. Huntford, *Scott and Amundsen*, (London: Hodder and Stoughton, 1979); K. Scott, *Self Portrait of an Artist*, (London: John Murray, 1949); L. Young, *A Great Task of Happiness; The Life of Kathleen Scott*, (London: Macmillan, 1995); G. Seaver, *Edward Wilson of the Antarctic*, (London: John Murray, 1933); G. Seaver, *'Birdie' Bowers of the Antarctic*, (London: John Murray, 1938); H.J.P. Arnold, *Herbert Ponting: Photographer of the World*, (London: Hutchinson, 1969); L.C. Bernacchi, *A Very Gallant Gentleman*, (London: Butterworth, 1933); S. Limb and P. Cordingley, *Captain Oates: Soldier and Explorer*, (London: Batsford, 1982); J. Lees-Milne, *Ancestral Voices*, (London: Chatto, 1975); A.F. Rogers, MD, 'The Death of Chief Petty Officer Evans, *The Practitioner*, April, 1974.

Mawson:

Mawson's Antarctic Diaries, ed. F. Jacka and E. Jacka, (Sydney: Allen and Unwin, 1988); Sir D. Mawson, *Home of the Blizzard*, 2 Vols., (London: Heinemann, 1915); Sir D. Mawson, *Home of the Blizzard*, Revd. Ed., (London:

Hodder and Stoughton, 1930); P. Mawson, *Mawson of the Antarctic*, (London: Longmans, Green, 1964); P. Mawson, *A Vision of Steel*, (Melbourne: Cheshire, 1958); L. Bickel, *This Accursed Land*, (Melbourne: Macmillan, 1977); C. Laseron, *South With Mawson* (London: Harrap, 1947); F. Hurley, *Argonauts of the South* (New York: Putnam, 1925); R. Huntford, *Shackleton*, (New York: Fawcett Columbine, 1985); J.G. Hunter, Unpublished Diary 1912–1913, Australian National Library MS.2806; R. Service, *Collected Poems*, (London: E. Benn, 1978); R. Dixon, *Writing the Colonial Adventure: Race, Gender and Nation in Anglo-Australian Popular Fiction, 1875–1914*; (Cambridge: Cambridge University Press, 1995).